PHILOSOPHY OF MATHEMATICS

Other books from Automatic Press ♦ $\frac{V}{I}$P

Formal Philosophy
edited by Vincent F. Hendricks & John Symons
November 2005

Thought$_2$Talk: A Crash Course in Reflection and Expression
by Vincent F. Hendricks
September 2006

Masses of Formal Philosophy
edited by Vincent F. Hendricks & John Symons
October 2006

Political Questions: 5 Questions for Political Philosophers
edited by Morten Ebbe Juul Nielsen
December 2006

Philosophy of Technology: 5 Questions
edited by Jan-Kyrre Berg Olsen & Evan Selinger
February 2007

Game Theory: 5 Questions
edited by Vincent F. Hendricks & Pelle Guldborg Hansen
April 2007

Legal Philosophy: 5 Questions
edited by Morten Ebbe Juul Nielsen
October 2007

Normative Ethics: 5 Questions
edited by Thomas S. Petersen & Jesper Ryberg
November 2007

Philosophy of Physics: 5 Questions
edited by Juan Ferret & John Symons
March 2008

Probability and Statistics: 5 Questions
edited by Alan Hájek & Vincent F. Hendricks
March 2008

PHILOSOPHY OF MATHEMATICS
5 QUESTIONS

edited by

Vincent F. Hendricks

Hannes Leitgeb

Automatic Press ♦ $\frac{V}{I}$P

Automatic Press ♦ $\frac{V}{I}$P

Information on this title: www.phil-math.org

© Automatic Press / VIP 2008

This publication is in copyright. Subject to statuary exception
and to the provisions of relevant collective licensing agreements,
no reproduction of any part may take place without
the written permission of the publisher.

First published 2008

Printed in the United States of America
and the United Kingdom

ISBN-10   87-99-1013-51   paperback

The publisher has no responsibilities for
the persistence or accuracy of URLs for external or
third party Internet Web sites referred to in this publication
and does not guarantee that any content on such
Web sites is, or will remain, accurate or appropriate.

Typeset in LaTeX2_ε
Graphic design by Vincent F. Hendricks

Contents

Preface	iii
Acknowledgements	v
1 Jeremy Avigad	1
2 Steve Awodey	11
3 John L. Bell	15
4 Johan van Benthem	29
5 Douglas Bridges	45
6 Charles S. Chihara	51
7 Mark Colyvan	75
8 E. Brian Davies	87
9 Michael Detlefsen	101
10 Solomon Feferman	115
11 Bob Hale	137
12 Geoffrey Hellman	145
13 Jaakko Hintikka	165
14 Thomas Jech	173
15 H. Jerome Keisler	175
16 Ulrich Kohlenbach	183
17 Penelope Maddy	191

18 Paolo Mancosu	193
19 Charles Parsons	205
20 Michael D. Resnik	211
21 Stewart Shapiro	219
22 Wilfried Sieg	233
23 William Tait	249
24 Albert Visser	265
25 Alan Weir	277
26 Philip Welch	289
27 Crispin Wright	301
28 Edward N. Zalta	313
About the Editors	329
About Philosophy of Mathematics	331
Index	332

Preface

Philosophy of Mathemtics: 5 Questions is a collection of interviews with some of the most influential scholars in the field of the last decades. We hear their views on philosophy of mathematics, its aim, scope, use, broader intellectual environment, the future direction of the philosophy of mathematics and how the work of the interviewees fits in these respects.

---------◆---------

Philosophy of mathematics is thriving without any advertising and thus the idea was not to produce a bill-board for philosophy of mathematics along the highways of science and buildings of academia. The idea was rather to produce a forum in which the giants of the field could speak their minds freely without the standard constraints of scientific severity imposed. Posing five open, relatively broad and even vague questions turned out to be the perfect format for this. Besides fascinating intellectual biographies, career recaps and racy memoirs, the contributions – either in the form of direct answers to the questions or as complete essays based on them – offer interesting new conjectures, pointers to work overlooked, and methodological reflections on mathematics and philosophy.

Given the format of this collection the purpose is not to articulate or push any particular agenda for philosophy of mathematics. We have also refrained from commenting on the responses to the questions. The responses are self-contained and readable, and no overarching view of philosophy and mathematics is lurking in the wings. The ambition is much more modest: to initialize the discussion as to how philosophers of mathematics understand their own enterprise and why these luminaries decide to make the intersection of philosophy and mathematics central to their work.

Admittedly, the decision to pursue this kind of work may in the end much be a matter of taste and personal inclination. However, in addition to containing succinct intellectual biographies the book does have some interesting, illuminating and most importantly *instructive* stories to tell about mathematics and philosophy and how they relate to each other. Since almost all of the central philosophical questions have counterparts in the philosophy of mathematics, the volume can be understood just as well as a panorama of analytic philosophy in the 20th century and beyond, looked at from one of its intellectual summits.

Vincent F. Hendricks & Hannes Leitgeb
Copenhagen and Bristol
December 2007

Acknowledgements

We are particularly grateful to the contributors for devoting time to writing such erudite, enlightening and often thought-provoking interviews and grateful to the philosophical community in general for showing interest in this project. In addition we would like to express our gratitude to Christopher M. Whalin and Elizabeth Pando for proof-reading and to Rasmus Rendsvig and Claus Festersen for assistance when encountering LaTeX-related problems. We would finally like to thank our publisher Automatic Press ♦ $\frac{\vee}{\mid}$P, in particular senior publishing editor V.J. Menshy, for continuing to take on these 'rather unusual academic' projects.

Vincent F. Hendricks & Hannes Leitgeb
Copenhagen and Bristol
December 2007

1

Jeremy Avigad

Associate Professor of Philosophy

Carnegie Mellon University, USA

Why I am a logician

In 1977, when I was nine years old, Doubleday released *Asimov on Numbers*, a collection of essays that had first appeared in Isaac Asimov's *Science Fiction and Fantasy* column. My mother, recognizing my penchant for science fiction and mathematics, bought me a copy as soon as it hit the bookstores. The essays covered topics such as number systems, combinatorial curiosities, imaginary numbers, and π. I was especially taken, however, by an essay titled "Varieties of the infinite," which included a photograph and short biographical sketch of Georg Cantor, and a brief introduction to his theory of the infinite.

Not long after, for my tenth birthday, my parents gave me a calculator. I had requested Texas Instruments' model TI-30, but my father surprised me with an SR-40, which had essentially the same functionality but came with a rechargeable battery. I did not know what all the buttons did, but I soon learned, from the manual, how to convert degrees to radians, and how to calculate the height of a flagpole from the length of its shadow and angle of inclination at the shadow's tip. This calculator instantly became one of my favorite toys.

What struck me about mathematics, at this early age, was this strange juxtaposition of fantasy and hard-nosed reality. It wasn't just that wild notions of the infinite and no-nonsense calculation could coexist within the same discipline; it was that they could be seen as complementary aspects of one and the same thing. When I was in elementary school and high school, "no-nonsense calculation" meant, for the most part, being able to write computer games with fancy graphics. But, at the same time, I was generally aware of the importance of mathematics to pursuits like economic forecasting and aeronautical engineering. As an undergraduate at Harvard, I majored in mathematics, and took the requisite core of courses in analysis, algebra, and geometry. But I also took a number of courses in computer science, including courses in data structures and algorithms, computational complexity, and graphics. Working part-time during the year and full-time over the summers, I wrote data analysis and control software for physicists, which served to bolster my faith in the links between abstraction and application.

I applied to graduate school in mathematics without a clear sense of what I wanted to do, beyond a vague desire to understand mathematics in computational terms and vice-versa. In my first semester at the University of California at Berkeley, I took Jack Silver's introductory logic course, and I was impressed by the promising blend of syntax and semantics. At the end of the year, Silver presented a model-theoretic proof of the Paris-Harrington theorem, which shows that a certain combinatorial statement is independent of first-order Peano arithmetic. Over the summer, I read a paper by Wilfried Buchholz and Stan Wainer bounding the rates of growth of the computable functions that can be shown to total in Peano arithmetic, and Robert Solovay and Jussi Ketonen's proof that witnesses to the Paris-Harrington statement cannot be so bounded. This was what I had been looking for: the place where lofty assumptions about the nature of the infinite come to bear on calculation. I decided to become a logician, and have never regretted that choice.

My use of the term "logician" here is deliberate. My interests lie in mathematics, philosophy, and computer science, and I have found institutional pressures to pigeonhole research under departmental labels to have had baleful effects on these subjects. Below, I will use the terms "mathematical logic" and "philosophy of mathematics" to describe sustained reflection on the goals and methods of mathematics, regardless of its disciplinary origin.

As will become clear, I identify myself most distinctly as a proof theorist, working in the tradition of Hilbert's program. The perspective that emerges from that program involves, first, modeling mathematical methods of argumentation in syntactic, axiomatic terms, and, second, using explicit, finitary methods to study the resulting notions of provability. Below, I will focus on the ways that such a modeling contributes to our philosophical understanding.

What we have learned

Popular presentations of the history of modern logic make much of the discovery of set-theoretic paradoxes at the turn of the century, Brouwer's intuitionistic challenge, the ensuing crisis of foundations, Hilbert's attempt to settle the question of foundations once and for all, and the devastating effects of Gödel's incompleteness theorems. This dramatic arc makes for a good story, but one that ultimately lacks depth, and renders the episode little more than light entertainment for contemporary mathematicians who do not lose sleep over whether their methods are consistent.

The real drama emerges in a broader historical context. Late nineteenth century developments brought radical changes to the very conception of what it means to do mathematics, and we are still living with the aftermath. It is a remarkable fact that work of Euclid, Descartes, Newton, Leibniz, Euler, and Gauss can still be read, profitably and with great enjoyment, today. But although we recognize their work as mathematics *par excellence*, it is mathematics of a certain character, with a more explicit computational flavor than is now common. It was not until the turn of the twentieth century that mathematicians began to view their subject as a general theory of structures, often characterized in set-theoretic terms, rather than as a theory of calculation, involving the symbolic expressions that are now taken to denote elements of the abstract structures. The set-theoretic paradoxes were only one pointed manifestation of a general worry that mathematics was taking a wrong turn; the broader concern was whether the new methods, even if consistent, were meaningful, and appropriate to the subject. Today, mathematicians strive for conceptual understanding and powerful methods of calculation, and these two aims sometimes exert distinct methodological pressures. My goal, as a logician and philosopher of mathematics, is to understand how mathematics manages to balance the two.

Too many philosophers today have the annoying habit of commending themselves for having the audacity to take on questions that are so hard as to be, perhaps, ultimately unanswerable. I can't imagine a poorer excuse for failure to make progress with one's research, and I worry that when the public at large begins to tire of such talk, the field will suffer a severe backlash. The contrapositive to the claim that philosophy deals with unanswerable questions is that in those happy situations where we find ourselves with satisfying answers to fundamental conceptual questions, neither the questions nor the answers can be counted as properly philosophical. That attitude, too, is unfortunate, because it prevents the field from taking credit for some important advances. In this section, I'd like to enumerate, briefly, some of the ways that mathematical logic and the philosophy of mathematics have contributed to our understanding of the methodological tension I have just described.

The syntactic, axiomatic standpoint has enabled us to fashion formal representations of various foundational stances, and we now have informative descriptions of the types of reasoning that are justified on finitist, predicative, constructive, intuitionistic, structuralist, and classical grounds. The positions can be cast in different ontological terms, for example, in the language of set theory, first- or higher-order arithmetic, or type theory. By now, we understand these frameworks quite well. We have come to learn what sorts of things can and can't be done in the various foundational systems, and we have discovered a number of interesting relationships between them.

One of the things we have learned over the course of the twentieth century is that much of mathematics does not need strong set-theoretic axioms. Of course, a lot hinges on the interpretation of the word "need" in this assertion. Set-theoretic language is now pervasive, and few mathematicians have any qualms about quantifying over uncountably infinite domains. But although the general principles that are invoked are, indeed, logically strong, this strength can typically be shown to be avoidable when one focuses on the particular ways in which the principles are used. In other words, foundational research has shown that one can generally describe workable formalizations of common mathematical developments in comparatively weak axiomatic theories. This sort of analysis tells us two things: first, that ordinary mathematical methods often have more constructive content than is immediately apparent, and, second, that logical strength alone does not account for all the benefits of modern methods, whatever these may be.

Throughout my career, I have been particularly interested in the ways that ordinary mathematical arguments can be understood in computational terms. Axiomatic theories can be used to model not just foundational stances, but also general patterns of reasoning, such as induction, compactness, and set existence principles like comprehension and choice. In studying such theories, I have made use of all the tools of the trade, including cut elimination, double-negation interpretations, realizability interpretations, and model-theoretic techniques. These provide ways of "reducing" classical theories to constructive ones, and "extracting" the computational and combinatorial content from nonconstructive arguments.

This work stands firmly in the proof-theoretic tradition. I do not have enough space here to provide an overview of the field, but I am pleased to see that many of those whose work I most admire have been asked to write for this collection. Some of the things I have done that I am especially proud of involve understanding how semantic arguments in logic can be understood in syntactic terms, and vice-versa; using forcing methods in various ways in service of proof-theoretic goals; and finding ways of representing infinitary mathematical methods in surprisingly weak theories. More generally, I have tried as best I can to sustain the core values of the proof-theoretic tradition, while at the same time addressing topics of contemporary concern.

What we are learning

The interdisciplinary and fundamentally reflective nature of mathematical logic is its greatest asset, but also its greatest burden. The field is often viewed as too mathematical to be philosophy, too philosophical to be mathematics, and too rarefied in both senses to find a home in computer science. Logicians, forced to justify their activities to deans and funding agencies, generally resort to one of two strategies: either we emphasize the intrinsic interest and importance of our work, or we emphasize the applications to other branches of mathematics and computer science.

Both strategies are reasonable, but also dangerous. Locating the importance of a logical development in its applications to a branch of mathematics, X, only passes the buck. One has to further explain why X, in turn, is so important, and then, given that it *is* so important, why one should not just forget about logic and focus on X instead. On the other hand, appeals to "intrinsic interest" fall flat when the interest is recognized by only a dozen or so logicians working in an isolated technical corner of the field, and otherwise opaque to the mathematicians whose basic methods and concepts are supposedly being illuminated.

The best work in mathematical logic arises when the two standards of success are met simultaneously, that is, when we are left with the impression that we have learned something fundamentally important and interesting about the nature of mathematics, and when that understanding leaves us better equipped to engage in our mathematical practices. Many proof-theoretic developments of the twentieth century have that character: syntactic modeling of mathematical methods and concepts not only helps us understand them better, but also supports algorithmic and computational efforts that rely on such models.

But the axiomatic studies of the twentieth century are only a start. An ordinary mathematician may prefer an elementary argument over a more conceptual one, for example, since the elementary argument provides explicit computational information. Telling that mathematician that the conceptual argument can be formalized in a weak subsystem of second-order arithmetic, and that our metatheorems tell us that such explicit information can always be extracted in principle, won't get you very far. Nobody cares much about what can be done "in principle"; logic is only interesting insofar as it tells us something about the mathematics that we actually practice. We therefore need to continue to refine our metamathematical modeling, remaining mindful of the foundational, conceptual, and methodological questions that make that modeling worthwhile.

I once heard Natarjan Shankar, a computer scientist, declare that we are in the "golden age of metamathematics." I like that phrase, and find it apt. It took a number of decades for the basic conceptual apparatus and terminology of mathematical logic to settle, and now, a century later, we have a powerful set of analytic tools at our disposal.

There are a number of ways that these tools can be put to good use. For one thing, proof-theoretic methods can be used to find useful information hidden in ordinary mathematical arguments. Georg Kreisel described the process of extracting such information as "unwinding proofs," and Ulrich Kohlenbach has more recently adopted the term "proof mining." The idea is simple. The proof-theoretic reductions described in the last section show that classical methods can often be interpreted in constructive, finitary, or otherwise explicit terms. Therefore, one need only apply the methods to particular proofs, and then harvest the epistemological gains. In reality, things are more complicated. Ordinary mathematical proofs are not presented in formal systems, so there are choices to be made in the formal modeling. In addition, the general metamathematical tools have to be tailored and adjusted to yield the information that is sought. But Kohlenbach and his students have had a number of striking successes in fields like numerical analysis and fixed-point theory, and researchers like Thierry

Coquand have obtained interesting results by applying general insights from constructive mathematics to particular problems in algebra. I have been working with Philipp Gerhardy and Henry Towsner to apply similar methods in the realm of ergodic theory and combinatorics. The field is young, however, and it is only beginning to attract the attention it deserves from the logical and mathematical communities.

Another place where proof-theoretic methods and insights can be put to use is in the field of automated reasoning and formal verification. Since the early twentieth century, it has been understood that ordinary mathematical arguments can be represented in formal axiomatic theories, at least in principle. The complexity involved in even the most basic mathematical arguments, however, rendered most formalization infeasible in practice. The advent of computational proof assistants has begin to change this, making it possible to formalize increasingly complex mathematical proofs. Such methods are now being used to verify that descriptions of hardware and software components meet their specifications, something that is especially important when lives depend on their proper functioning. But the methods can also be used for the more traditional task of verifying ordinary mathematical proofs, and are especially pertinent to cases where proofs rely on computation that is too extensive to check by hand.

In 2004, I earned a measure of recognition from the formal verification community by verifying a proof of the prime number theorem, with the help of some students at Carnegie Mellon. Soon after, Georges Gonthier announced a verification of the four color theorem, and Thomas Hales announced a verification of the Jordan curve theorem. Hales has moreover launched an ambitious project to verify his proof of the Kepler conjecture, and Gonthier is currently overseeing a project to verify the Feit-Thompson theorem. Success in these endeavors requires, among other things, having the computer be able to verify straightforward mathematical inferences, without requiring the user to spell out the details. Getting computers to do so, in turn, requires a logical modeling and classification of the relevant inferences, and the development of procedures that can fill in the details automatically.

These are just some of the ways in which a proof-theoretic understanding can support computational developments in mathematics. Fields like computational algebra, computational number theory, computational ergodic theory, and so on, are persistent attempts to retrofit computational significance to mathematical developments that have developed free from computational concerns. A proof-theoretic perspective should be especially helpful when the computations themselves involve propositional data, in the tradition of Leibniz' "calculemus!" For example, in computer algebra systems, proof-theoretic methods can be used to manage constraints that can be used to simplify complex expressions. Syntactic methods of quantifier elimination play a key role in real algebraic geometry, where one wishes to manipulate sets and functions that can be described with prescribed linguistic resources. Quantifier elimination has also played an important role in the study of valued fields, and is a key component in theories of motivic integration. Within logic, the study of definability is often relegated to model theory and descriptive set theory, but computational concerns encourage a proof-theoretic perspective as well.

Yet another domain where a syntactic, foundational perspective is important is in the search for natural combinatorial independences, that is, interesting combinatorial principles that are independent of conventional mathematical methods. Harvey Friedman, in particular, has long sought to find exotic combinatorial behavior in familiar mathematical settings. Such work gives us glimpes into what goes on just beyond ordinary patterns of mathematical reasoning, and yields interesting mathematics as well.

What we have yet to learn

I have tried to show that mathematical logic has a lot to contribute to mathematics. But is that philosophy? Writing down axioms doesn't provide a sense in which these axioms provide *bona fide* knowledge, or tell us what the basic terms of the discourse refer to. Isn't that what philosophy is supposed to do?

Let us not downplay our philosophical gains. The axiomatic method clarifies the ontological commitments that are presupposed in mathematical developments, and tells us what can and cannot be done on the basis of those commitments. Proof-theoretic analysis also yields satisfying philosophical explanations as to how abstract, infinitary assumptions have bearing on computational concerns, and provides senses in which infinitary methods can be seen to have finitary content. Even if what you really want is a theory of mathematical knowledge and justification, it is hard to see how one can justify any sort of methods without a meaningful articulation of what these methods are supposed to accomplish, and an analysis of how the methods are suited to the goals. We do not need fairy tales about numbers and triangles prancing about in the realm of abstracta. What we need is serious thought about what it means to do mathematics, and why we do it.

In *The Problems of Philosophy*, Bertrand Russell noted that whenever philosophical inquiry reaches a stage where the questions can be posed with sufficient clarity to admit precise answers, then the inquiry is commonly viewed as scientific rather than philosophical. There may be some merit to this distinction: some of the most interesting philosophical developments arise in situations where we find that the conceptual resources available do not give us a sufficient grip on questions that force themselves upon us. In this section, I will focus on such situations.

To start with, we need more robust and informative models of mathematical proof. Coming on the heels of my praise of the axiomatic method, this claim may seem surprising. But one need not deny the successes we have had with the methods of contemporary logic in order to recognize that there are questions that these methods were not designed to address. One particularly salient one is the problem of multiple proofs: on the traditional logical story, the purpose of a proof is simply to warrant the truth of the resulting theorem, leaving it utterly mysterious why it is often the case that we are pleased to find a new proof of the same theorem. Of course, any mathematician will tell you that we learn different things from different proofs. So the question is by no means "unanswerable." On the contrary, we have lots of good ideas about how the answer should go. My point is simply that our logical theories currently have little to say about the matter, and that it will take honest philosophical work to fashion our intuitions into something precise.

Much of the difficulty stems from the fact that there is a wide gap between ordinary mathematical proofs and the logician's formal derivations. I have argued, in an essay titled "Mathematical method and proof," that although formal axiomatic systems provide good normative and descriptive accounts of the standards by which we judge proofs to be correct, the models are not rich enough to support other types of common evaluations. For example, we often talk about different styles of proof. Some proofs are algebraic, while others are geometric; some are abstract, while others are more concrete; some are explicitly computational, while others are more structural, or "conceptual"; and so on. It is difficult to explicate these characteristics with the formal axiomatic model, on which every proof can be viewed as a derivation in axiomatic set theory. It is not just a matter of taking definitions seriously, and noticing whether a proof uses the terms "complex number" or "algebraic variety." What really matters is the theoretical scaffolding, the conceptual resources at play, the ways that problems and goals are posed, and the general methods of analysis. It is these fuzzier aspects of proof that are ripe for logical and philosophical analysis.

The questions we ask need not always be cast in terms of proof. Mathematicians also solve problems, build theories, pose problems, formulate conjectures, introduce concepts, and model empirical data. Understanding mathematical proof will require understanding these types of activities as well. For example, making sense of the ways that we value certain mathematical proofs requires a better sense of why we value the resulting theorems, and making sense of the roles that concepts play in a mathematical proof requires a better sense of the roles that concepts play more generally.

Such reflection is germane to the "scientific" research I described in the last section. In proof mining, one has to manipulate ordinary mathematical proofs, with sensitivity to the types of information that is commonly sought. In formal verification, ordinary mathematical proofs are, again, the focus of attention. The goal of research in combinatorial independences is to find logically strong principles that are "natural" and "interesting." Such research can therefore be supported by a better understanding of ordinary mathematical practice, and, conversely, more faithful modeling of that practice can provide a better understanding of mathematical epistemology.

Every time I write about these topics, I feel compelled to conclude with a rallying cry and a plea for more troops. Mathematics is a fascinating subject, and there is a wealth of insight to be gained from disciplined reflection on its methods. But the fundamental questions are too broad, and the subject too deep, for a handful of heterodox logicians to carry the day. This is a vast undertaking, and there is great progress to be made. Please, join the cause.

2

Steve Awodey

Associate Professor of Philosophy

Carnegie Mellon University, USA

Why were you initially drawn to the foundations of mathematics and/or the philosophy of mathematics?

As a student of Mathematics and Philosophy in Marburg, Germany, I was fascinated by the tension between the philosophical accounts of logic and mathematics given by Frege, Russell, Wittgenstein and logical empiricists like Carnap and, on the one hand, the character of modern mathematics in fields like algebraic topology, as well as the constructive logic that was popular in Marburg. Trying to understand and reconcile these conflicting views forced me to think hard about each of them and ponder their relative merits and compatibility.

When I later studied at Chicago, I began to see how some, if not all, of the conflicts could be reconciled. The key question for me has always been: what makes Mathematics true? My side interest in Carnap, which I pursued with Linsky and Stein at Chicago, derives from Carnap's attempts to answer this question, and Tait's views about constructivism, mathematical truth, and proof certainly clarified things and helped me focus my thoughts. But the real breakthrough for me was learning category theory and studying with Mac Lane. He had a very high perspective on Mathematics and had also thought seriously about its relation to logic and other topics in philosophy of math, like structuralism. I think his views are grossly underestimated by most philosophers of mathematics, by the way.

My own thinking was very much advanced by learning category theory, which has strong connections to, and provides a very different and fresh perspective on, topics such as constructive logic, logicism, formalism, and structuralism. The questions about the relationship between logic and mathematics and the nature of mathematical truth that once vexed me seem much more tractable from the vantage point provided by category theory.

Of course, my own mathematical research in logic and category theory actually occupies far more of my time and energy than such philosophical questions; but this "philosophical motivation," as it were, has definitely played a role in determining the direction of my mathematical interests.

What examples from your work (or the work of others) illustrate the use of mathematics for philosophy?

Questions of logical (in-)completeness, particularly of higher-order systems, are of great interests to some philosophers of logic and mathematics, and category theory provides some insights not available from other sources. I've proved some results in this direction myself, and tried to explain them informally in e.g. my paper "Continuity and logical completeness". Specifically, if one considers not just standard models but also ones that vary continuously over a topological space (as is done in topology with the notion of a "sheaf of groups"), then higher-order logic is actually deductively complete with respect to such models. So Gödel incompleteness marks a difference between the logic of constant models and that of variable ones.

The debate about structuralism in the philosophy of mathematics just cries out for a bit of category theory, as I tried to explain in my note "An answer to G. Hellman". And, more generally, I think a new philosophical position regarding mathematical knowledge that is inspired by, or at least compatible with, Carnap's later views can also be built around certain insights provided by category theory. I've tried to sketch it in the final section of my contribution to the recent *Cambridge Companion to Carnap*, where I called it "structural empiricism" — just for the fun of getting to coin a new "-ism"!

What is the proper role of philosophy of mathematics in relation to logic, foundations of mathematics, the traditional core areas of mathematics, and science?

I suppose this goes both ways:

First, philosophers of mathematics clearly need to look more closely at the actual practice of mathematics (today, not 100 years ago). This seems to be becoming more clear nowadays, after an embarrassingly scholastic period in the field of late.

But that's not to say that philosophical reflection has nothing to offer mathematics itself—on the contrary, I think that it can and does promote better, deeper, and more important mathematics. But that's only going to happen if the philosophical insights are themselves mathematically informed in the first place. A good modern example of this, I think, is in the work of F.W. Lawvere, who has made some very deep contributions to logic and philosophy of mathematics on the basis of insights gained through mathematics and category theory, and conversely. More famously, David Hilbert—and Saunders Mac Lane, I think—both made very deep contributions on both the philosophical and mathematical sides, partly as a result of their reflections in both areas.

What do you consider the most neglected topics and/or contributions in late 20th century philosophy of mathematics?

Category theory.

What are the most important open problems in the philosophy of mathematics and what are the prospects for progress?

The single most important open problem in current philosophy of mathematics in my view is coming to terms with the development of mathematics that has occurred in the last century. In that time, while philosophers were preoccupied with ontological commitments and Gödel incompleteness, mathematics itself has exploded with new methods, problems, fields of study, foundations, applications, and even style of exposition, none of which was even noticed by most philosophers of mathematics. It seems that even some contemporary philosophers still think that mathematics is essentially axiomatic set theory or some other formal deductive system (not that that was ever true).

As for progress in philosophy of mathematics, none will be made until philosophers actually look at mathematics (or mathematicians get more involved in philosophy). Once they do, however, I think the prospects are actually quite bright. Thanks in part to the wide-spread use of tools like category theory, the degree of coherence and level of uniform rigor in modern mathematics is now very high, making it possible to recognize certain features and aspects that were once obscured (like the role of axioms). It seems to me entirely plausible that, now that it has come into focus, we are finally in a position to achieve an understanding of some of the most basic philosophical questions, about the nature of mathematics, and our knowledge of it.

References

Awodey, S.: "An Answer to G. Hellman's Question 'Does Category Theory Provide a Framework for Mathematical Structuralism?"' *Philosophia Mathematica* (3), vol. 12, pp. 54–64, 2004.

Awodey, S.: *Category Theory*. Oxford Logic Guides 49, Oxford University Press, 2006.

Awodey, S.: "Continuity and Logical Completeness: An Application of Topos Theory." In: *The Age of Alternative Logics*, G. Heinzmann (ed.), Kluwer, 2006.

Awodey, S.: "Carnap's Quest for Analyticity: The *Studies in Semantics*."To appear in: *The Cambridge Companion to Carnap*, R. Creath & M. Friedman (ed.s), Cambridge University Press, (forthcoming).

Awodey, S. & C. Butz: "Topological Completeness for Higher-Order Logic." *Journal of Symbolic Logic* 65(3), pp. 1168–82, 2000.

3
John L. Bell

Professor of Philosophy

University of Western Ontario, Canada

Why were you initially drawn to the foundations of mathematics and/or the philosophy of mathematics?

My route to the foundations and philosophy of mathematics was somewhat circuitous. In youth I was attracted to physics, especially relativity theory and cosmology—I actually attended one of Fred Hoyle's lecture courses on the subject in Cambridge in the early 1960s. (Parenthetically, I may mention that it was through Hoyle's lectures that I first heard the name Gödel, not of course in connection with his discoveries in logic, of which I was then wholly ignorant, but as the deviser of cosmological models containing closed timelike lines.) While I was, I suppose, quite clever at solving problems in mathematical physics and analysis – as I still joke, I could raise and lower a tensor index with the best of 'em! – after a while I began to realize that I had no genuine understanding of what I was actually doing. In particular, I was not even sure what a tensor really *was*. At the risk of joining the fabled centipede whose effort to understand its mode of locomotion reduced it to complete immobility, I decided to turn away from physics, my first love, and concentrate on pure mathematics. While mathematics lacked, in my eyes, the romantic appeal of cosmology, it had the compensating merit that its concepts and methods could, in principle at least, be fully presented to the understanding. My flight to mathematics was fuelled by my discovery of John Kelley's classic work *General Topology*. Its unique combination of mathematical elegance and dry wit, together with its extraordinary collection of exercises, stimulating but never oppressive, made a big impact on me. In particular, I was intrigued by the series of exercises on Boolean algebras (rings, but no matter) which I attempted to work through. Kelley also furnished my first introduction to set theory. Reading Kelley led me to study Gödel's monograph, *The Consistency of the Axiom of Choice and the Generalized Continuum Hypothesis.* The first two-thirds of this mathematical tour-de-force, in which Gödel presents his axiom system for set theory and develops its essential properties, seemed reasonably clear. But, despite my best efforts, I was unable to fathom the final part of the work, its grand finale, so to speak, in which, accompanied by an inaudible clash of cymbals, the consistency of the GCH is established. A good few years were to pass before I felt I truly understood what was going on.

Another influence was Bourbaki's *Éléments de Mathématique.* On first coming across some volumes of this monumental work in Blackwell's bookshop, I was excited to find that it was intended to be a complete, systematic account of abstract mathematics, precisely the kind of mathematics to which I had already been converted by Kelley's *General Topology.* The *oeuvre Bourbachique* included not only *Topologie Génerale,* but *Algèbre, Thèorie des Ensembles, Espaces Vectoriels Topologiques, Algèbre Commutatif*—magical titles in my eyes. I bought as many volumes as I could afford, often in obsolete – and so cheaper – editions (the whole enterprise seemed to be undergoing constant revision), and commenced to work my way through the collections of challenging exercises at the end of each section. I toiled mightily, in particular, to formulate solutions to the exercises on ordered sets in Chapter 3 of the *Thèorie des Ensembles.* It was from these that I first learned about ordinals, which Bourbaki presents in the original Cantorian manner as order types of well-ordered sets.

Kelley, Bourbaki, Gödel: it was through their influence that I was led to the foundations of mathematics. My interest in philosophy, on the other hand, derived from my being a voracious and eclectic reader. As an undergraduate I recall reading Plato's *The Last Days of Socrates,* William James's "Essays on Pragmatism", G.E. Moore's philosophical essays, Hegel's *Philosophy of History,* Descartes' *Discourse on Method,* Spinoza's *Ethics* (the statements of the theorems at least, since I found the "proofs" unenlightening), Leibniz's delphic *Monadology,* some Locke, Berkeley and Hume, Schopenhauer's *Essays in Pessimism.* And of course Bertrand Russell's breezily brilliant, if irresponsible, *History of Western Philosophy.* My attempts to penetrate the profundities of Kant's *Critique of Pure Reason* were frustrated by the work's apparent indigestibility. (I was only to appreciate its depth and philosophical importance many years later.) I greatly enjoyed Hans Reichenbach's *Philosophy of Space and Time*. On Blackwell's shelves in Oxford I came across Norman Malcolm's *Wittgenstein: A Memoir*. I was deeply moved by Malcolm's portrayal of Wittgenstein, in which he emerges as an intellectual ascetic of compelling moral grandeur. Wittgenstein's tiniest defiances of convention, for example, his refusal to wear a tie at dinner in Trinity College, I found admirable. Reading Malcolm's memoir stimulated me to attempt to read Wittgenstein's philosophical works. I was intrigued by the *Tractatus Logico-Philosophicus*, a masterpiece of sybilline refinement and compression in which Wittgenstein embarks on the heroic effort of reducing philosophy to the expressible, but in the end washes up on the shores of the ineffable. The conventionalism of the later Wittgenstein's *Philosophical Investigations*, I found less appealing.

I was drawn quite early on to the foundations of mathematics, and to general philosophy, but my conscious interest in the philosophy of mathematics per se was comparatively slow to crystallize. This took place in three stages. First, as an undergraduate I had developed an interest in set theory and the philosophy of the infinite in general. Next, I was a product of the politically supercharged 1960s, a time in which many young people, myself included, began to think about the social and political implications of their own activity, which in my case was the practice of mathematics (by this time mathematical logic). I became a member of a group of like-minded *gauchiste* mathematical logicians determined to terminate the funding of logic conferences by military-imperialistic sources such as NATO. All of us, I think I may safely say, believed that mathematics, while being like art, a beautiful sublimation of human activity, has, in the final analysis, to be understood as the product of actual human beings living in the world. Such stimuli led me (for better or worse) to see that mathematics actually has a *hidden content*, which can actually be *argued about*. This is the opposite of the unthinking Platonism/realism to which I was, I guess, initially attracted as offering the simplest account of mathematical truth, and which also possessed the additional advantage of avoiding what I then felt to be a certain cynicism inherent in Formalism. (Still, as I have come to learn, Formalism has the great merit of offering the weary ex-Platonist a refuge.) But, like the child's loss of belief in Santa Claus, I came to regard the Platonistic account of mathematical entities as a kind of fairy tale, and in any case as engendering insuperable epistemological difficulties. I may parenthetically remark that I have since come to liken Platonism to a (necessary) disease, which, like measles, must have been contracted in one's youth so as to confer an immunity in later life.

The third stage in the development of my interest in the philosophy of mathematics came through my efforts to understand *topos theory*. I was very struck by Bill Lawvere's insight that a topos is an objective presentation of the idea of *variability*, and that its internal – intuitionistic – logic may be considered as a logic of variation. Later I went so far as to attempt to use the topos concept as the basis for a "local" (as opposed to "absolute") interpretation of mathematical statements. I suggested that the unique absolute universe of sets central to the orthodox set-theoretic account of the foundations of mathematics should be replaced by a plurality of local mathematical frameworks – elementary toposes – defined in category-theoretic terms. Observing that such frameworks possess sufficiently rich internal structure to enable mathematical concepts and assertions to be interpreted within them, I maintained that they can serve as local surrogates for the usual "absolute" universe of sets. On this account mathematical concepts will in general no longer possess absolute meaning, nor mathematical assertions (e.g. the continuum hypothesis) absolute truth values, but will instead possess such meanings or truth values only locally, i.e., relative to local frameworks. The absolute truth of set-theoretical assertions would then, I held, give way to the subtler concept of invariance, that is, validity in all local frameworks. Thus, e.g., while the theorems of constructive arithmetic turn out to possess the property of invariance, the axiom of choice or the continuum hypothesis do not, because they hold true in some local frameworks but not others.

I still find this view attractive, but it is, after all, only one among many possible accounts of mathematics. If I were pressed to characterize my present attitude towards the foundations of mathematics, I would use the word *pluralistic*: no unique foundation, rather an interlocking ensemble of "foundations".

What examples from your work (or the work of others) illustrate the use of mathematics for philosophy?

There are, of course, numerous examples illustrating the use of mathematics for philosophy. "Negative" examples include the Pythagorean discovery of incommensurable magnitudes, Zeno's paradoxes, and the Gödel incompleteness theorem, each of which served to refute a certain philosophical doctrine. "Positive" examples include the Pythagorean discovery of the arithmetical basis for harmony, their invention of figurate numbers, Euclidean geometry, the infinitesimal calculus, Riemannian geometry, set theory, probability theory, relativity theory, quantum theory, and the theory of computation, all of which have been important influences in shaping philosophical views.

One can find other, more specific, examples of such influence in the 19th and 20th centuries. Frege's work on the foundations of arithmetic (and Bolzano's before him) is now held to have contained the seeds of what was later to flower into analytic philosophy. Russell's philosophical views were profoundly influenced by his work in mathematical logic. Brouwer's philosophy of intuitionism was first and foremost a philosophy of mathematics. Well-known is the impact Tarski's theory of truth had on Popper's philosophical outlook, serving, as it did in the latter's eyes, to revive the correspondence theory of truth.

Another important source of interaction between mathematics and philosophy arises from the opposition between the continuous and the discrete. Synechism, the doctrine that the world is ultimately continuous, has been defended by the majority of philosophers in the past, including Aristotle, Descartes, and Kant. (Leibniz seems to wavered, ending up with his strange hybrid doctrine of monadism.) Atomism, the doctrine that the world is ultimately particulate, was for a long time considered a maverick position. Now however, owing primarily to work in physics and chemistry, and lately also to the emergence of computing machines, it appears to be gaining the upper hand. (The movement to reduce mathematics to set theory initiated in the 19th century can already be seen as a victory for a form of atomism.) Synechism and atomism, along with the various syntheses of the two that have emerged in the history of thought, have been developed primarily in *mathematical* terms.

I believe that a significant potential influence of mathematics on philosophy may be seen in *category theory*. Category theory arose as a general apparatus for dealing with mathematical structures and their mutual relations and transformations. From a philosophical standpoint, a category may be viewed as an explicit presentation of a *form or concept*. The objects of a category are the *instances* of the associated form and its morphisms or arrows are the transformations between these instances which in some specified sense "preserve" this form. Functors between categories may then be considered as embodiments of morphological variation—change of form. Category theory is beginning to be seen as an appropriate language for describing not just mathematics, but the world, in structuralist terms, in terms of form. On this account there is no unique category (or topos) representing the objective world, but a number of different categories each embodying an idealization of a significant feature of the world. For instance, the topos of sets embodies the idea of discreteness, the smooth topos that of continuity and differentiability, and the effective topos that of computability. Each topos possesses properties not shared by the others: in the topos of sets the axiom of choice (and hence classical logic) holds; in the smooth topos the real line is indecomposable; and in the effective topos the space of countable sequences of natural numbers is enumerable. Each of these features can be seen as a necessary consequence of the particular form of idealization involved.

After these spectacular instances of the impact of mathematics on philosophy, it comes as something of an anticlimax to mention, as suggested, some of my own modest contributions to that area. The first of these was essentially a contribution to philosophy of science. As an ex-aspiring-physicist I had long been intrigued by quantum theory, with its mysterious superpositions of states and incompatible measurements; and as a logician my curiosity was piqued by the so-called quantum logic, whose characteristic feature is that its algebra of propositions is not a Boolean or Heyting algebra, but a certain kind of nondistributive lattice—an ortholattice. All of these facts can be, and are, formally derived from the standard Hilbert space formalism of quantum theory. I became interested in the problem of formulating some simple principles, free of the technicalities of the theory of Hilbert spaces, from which one could derive the anomalous features of quantum theory, as well as the ortholattices underlying quantum logic. I came up with two approaches. The first, essentially topological, was based on the idea of using what I called a proximity space, a set equipped with a symmetric reflexive relation "close to". The lattice of parts of such a space is an ortholattice. There is a natural way, which I called "manifestation", similar to Paul Cohen's celebrated concept of set-theoretic forcing, of relating propositions (actually attributes) to parts of the space. The propositions manifested over the whole of every proximity space are (essentially) the theses of quantum logic. Given two propositions P, Q, their superposition can be identified with $\neg\neg(P \vee Q)$, and they are incompatible if there is a proximity space with a part manifesting P but not $Q \vee \neg Q$, or vice-versa.

In my other approach to the problem, I showed how to construct the ortholattices arising in quantum logic from what I saw as the phenomenologically plausible idea of a collection of ensembles subject to passing or failing various "tests". A collection of ensembles forms a certain kind of preorderd set with an additional relation I called an orthospace: I showed that the complete ortholattices, in particular those of quantum theory, arise as canonical completions of orthospaces in much the same way as arbitrary complete lattices arise as canonical completions of partially ordered sets. I also showed that the canonical completion of an orthospace of ensembles may be identified with the lattice of properties of the ensembles, thereby showing exactly why ortholattices arise in the analysis of "tests" or experimental propositions. I went on to axiomatize the concept of "test" itself in terms of the more primitive notion of "filters" acting on ensembles. "Passing" an ensemble through a filter s produces the subensemble of entities that have "passed" the test corresponding to s. Two filters s and t can be juxtaposed to produce the compound filter st, but in general $st \neq ts$. When this latter is that case, the two tests corresponding to s and t are, like position and momentum measurements in quantum theory, not simultaneously performable, that is, *incompatible*. When (and only when) $st \neq ts$, the juxtaposition of s and t corresponds to their logical conjunction. In this setting, it is the noncommutativity or incompatibility of filters or "tests" that gives rise to "quantum logic".

A philosophical problem that had long intrigued me was: why is traditional logic *bivalent*, that is, why is it assumed that there are just two truth values rather than some other number? What is it about the number 2 that gives it this special position in logic? Wittgenstein seems to take the fact for granted when (in his Notes on Logic) he says that propositions have two "poles". It is often claimed that bivalent logic is the "logic of realism", that is, logic in which propositions are construed as referring to independently existing objects, in contrast with "anti-realist" logics such as intuitionistic logic (I don't agree that intuitionistic logic has to be thought of as anti-realist—but let that pass). However, this begs the question, since the thought immediately arises: what is it about the realm of independently existing objects that confers bivalence on propositions referred to it? Why shouldn't the number of objective truth values be, say, 3, like the number of spatial dimensions? Wittgenstein recognized the possibility of this question arising but simply dismissed it.

One way that occurred to me of explaining the role of the number 2 in logic is by moving from individual propositions to sets of propositions, or *theories*. Frege had suggested that the bivalence of the logic of concepts arises from their having *sharp boundaries*: one can determine with exactitude, for such a concept, when an object falls under it, or when it does not. In other words, a concept's possession of a sharp boundary means that the theory of the concept is complete with regard to atomic propositions. It is then natural to extend this prescription to arbitrary propositions. So, metaphorically, we may say that (the concept determined by) a theory has sharp boundaries if it is *complete*, that is, if any proposition in the theory's vocabulary is provable or refutable from the theory. But it is well known that, for any complete theory T (in propositional intuitionistic or classical logic), it is possible to assign the *two* truth values 0, 1 to propositions in such a way as to respect the logical operations, and also to assign precisely the propositions in T the value 1. And conversely, if such a bivalent assignment exists, the theory is complete. That is, the number 2 is simply the numerical representative of completeness, or the possession of "sharp boundaries".

The major logical consequence of bivalence (although not equivalent to it) is the *law of excluded middle:* the assertion, for any proposition P, of the disjunction $P \vee \neg P$. This is of course the logical principle which whose affirmation distinguishes classical from intuitionistic logic. Like bivalence, the law of excluded middle has been taken to be characteristic of logic in which propositions are construed as referring to independently existing objects. I found that, if one starts with intuitionistic predicate logic, and extends it to include Hilbert's ε-terms (these are essentially objects named by the use of the indefinite article: *a* such-and-such), then the law of excluded middle becomes provable. That is, the law of excluded middle is, after all, derivable from what can reasonably be construed as an ontological principle.

I also found myself attracted by the recent revival of interest in Frege's attempt to derive arithmetic from logic, in particular to the central mathematical result, now known as *Frege's Theorem*, implicit in his *Grundlagen*. Stated in set-theoretic terms, Frege's Theorem reads: for any set E, if there exists a map ν from the power set of E to E satisfying the condition

$$\forall XY[\nu(X) = \nu(Y) \Leftrightarrow \text{there is a bijection between } X \text{ and } Y],$$

then E has a subset which is the domain of a model of Peano's axioms for the natural numbers. My first piece of work on Frege's theorem was to observe that it can be proved by the same means as Zermelo used to derive the well-ordering theorem from the axiom of choice. I then became interested in the question of whether Frege's theorem can be proved constructively. I found that this was indeed the case, providing a constructive proof of a "best possible" version of Frege's theorem in which the premise is weakened so as to require only that the map ν be defined on the family of *finite* subsets of the set E. I also showed that that the postulation of such a structure (E, ν) – a *Frege structure* – is constructively equivalent to the postulation of a model of Peano's axioms.

What is the proper role of philosophy of mathematics in relation to logic, foundations of mathematics, the traditional core areas of mathematics, and science?

I think that the philosophy of mathematics should, and in fact does, play a dialectical role in relation to its sister disciplines, guiding them and, reciprocally, responding to their internal development. Let me attempt to illustrate what I mean. Cantor's philosophy of the infinite (and his associated, if lesser-known, championship of the reduction of the continuous to the discrete) played a major part in his development of set theory, which, as is well-known, came to permeate mathematics. Partly in reaction to the unrestricted use of Cantorian set theory in mathematics, Brouwer formulated his philosophy of intuitionism which in *its* turn radically influenced his mathematical practice and that of his immediate followers, a practice which was to prove seminal for constructive and computational mathematics. In *its* turn the latter is generating its own philosophy ... And so it goes.

What do you consider the most neglected topics and/or contributions in late 20th century philosophy of mathematics?

In my view contemporary philosophers of mathematics (or at least those who can be described as "mainstream") have paid far too much attention to set theory, ignoring the philosophical import of other major developments in mathematics such as category theory, type theory, and constructive mathematics. The impressive– and they *are* impressive – achievements of set theory in advancing mathematical knowledge have, perhaps, (mis)led these philosophers into thinking that, as far as philosophy is concerned, mathematics just *is* set theory. (This is the same "mistake" I believe Russell made when he claimed that mathematics just *is* logic, only in his case the "mistake" had the positive – if, with hindsight, accidental – consequence of leading to type theory.) But in truth set theory represents only one side of an opposition – that between the continuous and the discrete – which is still stimulating the growth of mathematics. With the introduction of set theory, mathematics was reduced to pure discreteness (in the eyes of certain philosophers) and those aspects of continuity incompatible with discreteness (e.g. infinitesimals and indecomposable continua) were driven out. With the emergence of category theory, type theory and constructive mathematics, set theory, while still dominant, can now be seen as no more than one among a number of ways of depicting the mathematical universe. I believe that it would benefit philosophers of mathematics to become aware of this fact.

Another topic which I think has been, on the whole, neglected by contemporary philosophers of mathematics (there are, admittedly, exceptions) is the *applicability* of mathematics. Mathematics is perhaps unique in being at once art and science. As an art, it is free to develop aesthetically pleasing internal practices of its own, practices which are capable of reduction to simpler, but equally beautiful practices which can then function as rules. (The art with which it is natural to compare mathematics in this regard is music, in which the simple rules governing the diatonic scale came to serve as the "foundation" for musical composition.) But mathematics is also a *science*; it serves to describe the natural world – in the terms of idealist philosophy, a *transcendent* world – a world that exists independently of it. The correlation between the internal practice of mathematics and the properties of the natural world is remarkable and seems to demand some kind of explanation. Galileo's explanation was that mathematics was the language of "the book of nature"; but with the rise of quantum theory and other esoteric physical theories, couched in exotic mathematical terms, physicists have become less comfortable with this explanation. It seems almost a miracle, for example, that the mathematics of Hilbert space, invented for an entirely different purpose, serves perfectly to represent the mechanics of the microworld. (This and other such "coincidences" led the physicist Eugene Wigner to entitle a famous paper "The Unreasonable Effectiveness of Mathematics".) I believe that philosophers of mathematics should enlarge their program of explicating the internal workings of mathematics to embrace the connection between mathematics and the outer world.

What are the most important open problems in the philosophy of mathematics and what are the prospects for progress?

Here are some problems in the philosophy of mathematics which to me possess significance, and are unquestionably still "open".

To explicate the applicability of mathematics

I have discussed this in the previous section.

To understand how the brain/mind generates mathematical concepts

This seems to me a problem as perplexing and intriguing as that of how the brain "generates" consciousness. A most interesting attempt to grapple with this problem has been made by the linguist George Lakoff and the psychologist Rafael Nuñez. In their book *Where Mathematics Comes From*, they fashion a sophisticated naturalistic explanation of the origins of mathematics. They advance the thesis that mathematics is not the product of some mysterious synergy between the mind and some putative empyrean world. They contend that mathematics, in all its richness and elaboration, emerges through the natural interaction of the cognitive processes common to us all with our experience of living in the actual world. We are from birth equipped with certain rudimentary mathematical abilities, for instance, the ability to distinguish objects, that of grasping and comparing the size of small pluralities instantly, and that of adding and subtracting small whole numbers. It is Lakoff's and Nuñez's contention that mathematics has emerged from our modest initial cognitive endowments through the brain/mind's distillation of *conceptual metaphors* from its/our experience of the external world, e.g. that arising from bodily movement, physical force, and spatial orientation. In terms reminiscent of category theory, Lakoff and Nuñez define a metaphor as a correlation or mapping grasped by the mind between two conceptual domains, the first of which, the source domain, is relatively concrete and familiar, and the second, the target domain, is of a more abstract character. Like morphisms in a category, these correlations must preserve structure. Thus, for example, in the metaphorical correlation between, say, heaps of stones (the source domain) and numbers (the target domain), our grasp of the fact that the combination of two piles of stones each of a given size always results in another pile of stones of a certain related size is projected onto the domain of numbers as the operation of addition. According to Lakoff and Nuñez, metaphorical correspondences such as this are the *fons et origo* of mathematical thought. While their claim is, of course, unprovable in a literal sense, I like it both for what I see as its essential plausibility and for the fact that it addresses a problem that has always nagged me.

To explicate the relationship between the continuous and the discrete—in particular, to explain how, continuity emerges from a discrete world

I have already touched on this problem – the significance of which extends far beyond the philosophy of mathematics – in a number of places above.

Here let me mention what seems to me an important special case of the problem: how is the continuity of perception (that of vision, for example) engendered by a discrete system of receptors? Actual perceptual fields can be modelled by *proximity spaces*. A proximity space is a set S equipped with a *proximity relation*, that is, a symmetric reflexive binary relation $\approx$. Here we think of S as a field of perception, its points as *locations* in it, and the relation $\approx$ as representing *indiscernibility of locations*, so that $x \approx y$ means that x and y are "too close" to one another to be perceptually distinguished. Let us call a proximity space $(S, \approx)$ *continuous* if for any $x, y \in S$ there exist $z_1, ..., z_n$ such that $x \approx z_1, z_1 \approx z_2, ..., z_{n-1} \approx z_n, z_n \approx y$. Continuity in this sense means that any two points can be joined by a finite sequence of points, each of which is indistinguishable from its immediate predecessor. If d is a metric on S such that the metric space (S, d) is connected, then every proximity structure determined by d is continuous. When S is a perceptual field such as that of vision, the fact that it does not fall into separate parts means that it is connected as a metric space with the inherent metric. Accordingly every proximity structure on S determined by that metric is continuous. Note that this continuity emerges even when S is itself an assemblage of discrete "points". In this way continuity of perception could be produced by a discrete system of receptors.

To explicate the role of computability in mathematics

How, in particular, does the computational structure of a mathematical result reflect its content? What is the relationship between the content of a mathematical theorem and the length or complexity of its proof? In the case of spectacular recent mathematical achievements such as the proofs of the Fermat theorem and the Poincaré conjecture, the comprehensibility of the proposition proved and the complexity of its proof would seem to be in inverse relationship. This is to be contrasted, with, say, category theory, in which propositions and their proofs are virtually on an equal footing as regards intelligibility.

To characterize how mathematics as a formal/symbolic practice differs from a practice such as fiction

Of course both are language-based (despite Brouwer's contrary claim that mathematics is a "languageless activity"). But more particularly, mathematics resembles fiction in its systematic introduction of concepts such as numbers, circles, sets, etc., which are then *reified*, that is, treated as if they possessed independent existence—this is as true of constructive as of classical mathematics, by the way. In fiction, characters and events are treated, in accordance with Coleridge's "willing suspension of disbelief", as if they were real. Now one important difference between *classical* mathematics and the practice of fiction is that the reified concepts of the former, but not the latter, are treated as if their properties were fully determinate. For instance, it is accepted (I would surmise) by the majority of mathematicians that it is objectively determined whether the number $10^{10^{10}} + 3$ is prime or not—even if, as is likely, *we* shall never know the answer. But in the case of fiction the case is otherwise. Scholars may debate Shakespeare's identity, but the question of whether Hamlet's breeches was, say, green, lacks determinacy, indeed borders on the absurd, since no scrutiny of Shakespeare's play could reveal their colour. Here the play is indeed the thing!

By contrast, the manner in which reified objects are treated both in constructive and in structuralist/axiomatic mathematics (category theory, for example) bears a closer resemblance to fiction. Constructive mathematicians acknowledge that the concepts and devices of mathematics are *invented* or *constructed*, even if under such (objective) constraints as to make it seem plausible later to describe them as the products of *discovery*. While in constructive mathematics *finite* objects such as individual natural numbers are treated as if their (finitistic) properties were fully determinate, (potentially) *infinite* objects such as the set of natural numbers, numerical functions, and individual real numbers are treated in a manner similar to fictional characters in that their properties are taken to be open to further determinations. The same can be said of structuralist mathematics. Just as Sherlock Holmes or Philip Marlowe have been the protagonists of numerous sequels to those works in which they made their debuts, so in structuralist mathematics there are a number of different ways of spelling out the properties of, for example, the real number system—"sequels", as it were, to its original conception. Models have been constructed in which every function on the real numbers is continuous, and also models in which every such function is computable. Like the practice of fiction, structuralist mathematics is *pluralistic*. I think the analogies—and the differences—between mathematics and fiction deserve further investigation.

Let me conclude by saying that I believe the ultimate purpose of the philosophy of mathematics is to demystify mathematics while at the same time celebrating it.

4

Johan van Benthem

University Professor of Logic and
Professor of Philosophy
University of Amsterdam & Stanford University
The Netherlands & USA

Why were you initially drawn to the foundations of mathematics and/or the philosophy of mathematics?

I started out as a student of physics, hard-working, interested, but alas, not 'in love' with my subject. Then logic struck, and having become interested in this subject for various reasons – including the fascinating personality of my first teacher – I switched after my candidate's program, to take two master's degrees, in mathematics and in philosophy. The beauty of mathematics was clear to me at once, with the amazing power, surprising twists, and indeed the music, of abstract arguments. As our professor of Analysis wrote at the time in our study guide "Mathematics is about the delight in the purity of trains of thought", and old-fashioned though this phrasing sounded in the revolutionary 1960s, it did resonate with me. Then I had the privilege of being taught set-theoretic topology by a group of brilliant students around De Groot, our leading expert around the time, who worked with Moore's method of discovering a subject for oneself. Topology unfolded from a few definitions and examples to real theorems that we had to prove ourselves – and the take-home exam took sleepless nights, as it included proving some results from scratch which came from a recent dissertation (as it turned out later). Only at the very end did De Groot himself appear, to give one lecture on Tychonoff's Theorem where an application was made of the Axiom of Choice, a sacral act only to be performed by tenured full professors.

At the same time, I was learning how mathematical logic and mathematics formed a natural unity, through my evening reading of Nagel & Newman's little book *Gödel's Proof* (1957), which reinforced this mystery that the deepest insights are accessible through mathematical clarity. Also later on, I have often been influenced by books bought on recommendation from other students, not as course requirements from our teachers – such as Benacerraf & Putnam's *Philosophy of Mathematics*, van Heijenoort's *From Frege to Gödel*, or the little gem edited by Jaakko Hintikka called *The Philosophy of Mathematics*, replete with logic classics. And then formal courses backed this up. I have been taught by intuitionists like Heyting and Troelstra, by formalists like Curry, and I eventually wrote my Ph.D. thesis in mathematics with Martin Löb, a very purist proof theorist, and a gentleman from the Grand Old German School in mathematical logic and the foundations of mathematics. Of course – these were the revolutionary early seventies – I immediately rejected all intellectual views of all my professors out of hand. I turned to modal logic, semantics, and model theory. Löb was disappointed. Himself a Jewish refugee from Germany, his view was that proof theory was about deep structure, elegance and style, qualities he attributed to German and Japanese culture, whereas model theory was about fast images and muddling through, which he associated with Anglo-Saxon and French culture. (In his highly original view of history, this explained the core alliances in the Second World War.) My own desire for independence was helped to some extent by these and other curious features. For instance, Curry really believed in formalism and formula games, and did not interact with his audience. I once was the only student in the room, but he still lectured at a distance as if there were a crowd. The exam consisted in two hours of formal deductions in his office – he collected streams of computer print-out paper for this purpose – in pure implicational propositional calculus, with no appeals to the Deduction Theorem allowed. Halfway through the exam, a silver-haired lady knocked on the door, and watched me struggle with formulas and paper. She said: "Good afternoon. I am Mrs. Curry, and I want you to know that my husband is a kind man ... ." That gentle gesture did more for me than the Deduction Theorem could ever have. I will save my student observations of the other professors for my Memoirs, and move ahead now.

I find it hard to separate thinking about mathematics itself, mathematical logic, 'foundations', and the philosophy of mathematics. The sensibilities needed for one may be close to those needed for the other. I definitely found what I was looking for in the delights of proving significant results in the meta-theory of modal 'intensional' logic, and through developing 'modal correspondence theory', seeing patterns and connections with classical 'extensional' logic that were below the surface of received opinion at the time, only brought to light by mathematical analysis. But I also had the transcendent moments of despair one needs to experience, when thinking about layer on layer of meta-analysis in set theory, or the reversals of perspective when translating one theory or logic into another, and then back again. Nowhere a piece of solid ground! It was like the way I often feel when lying on a beach and looking at the blue sky overhead: an immediate stomach-twisting sense of falling off into the Universe. Where was the safe Archimedean Standpoint which would prevent one from falling into bottomless intellectual pits? I would now see these experiences as necessary intellectual initiation rites – and they were far more mind-blowing than the drugs also available in my student environment at that time. What finally attracted me was the esotherical sense of joining a sort of secret society of logicians who could develop formal theories and then emerge in broad daylight in any discipline: mathematics, computer science, philosophy, or theology. This fitted my general intellectual inclinations, then, as much as, and now.

Of course, there is also a sociological aspect of initiation. Years later, when I was appointed in Groningen, I had to speak at a no-nonsense mathematics colloquium which was famous for intimidating newcomers and outsiders. I decided at once that I would not do what most people did, and try to impress them with cramming as much technical mathematics into my talk as possible. Instead, without any pre-amble, I started by telling them that I would prove the Existence of God, and began writing exotic modal formulas for that without much explanation. This surprise opening went down extremely well. Of course, after that, I had also prepared a lot of heavy modal logic stuff with ultraproducts and model theory, but the main point was made. After the talk, I was taken to the office of the department head, and offered some hard liquor from an illegal stock he kept there in a secret drawer. I knew that I had passed the test.

What examples from your work (or the work of others) illustrate the use of mathematics for philosophy?

There are so many great names right at the interface, and so many well-documented cross-currents between the two fields, that I have little to add to that. In some ways, Plato's inscription 'Mèdeis Ageometrètos Eisito" on the Gate of the Academy said it all, and 'the rest is history'.

Apart from the great names, I myself have always admired the gentle but persuasive influence of my predecessor Evert Willem Beth, the first occupant of the Amsterdam logic chair in the philosophy department, a companion to Brouwer's in mathematics. For Beth, insights into the history of mathematics and then contemporary work on its foundations had immediate repercussions for philosophy, and beyond that, for modern society. His magnum opus *The Foundations of Mathematics* was not a purely technical book, but it also contains his views on what happened when our views of science were transformed through the abstract postulational revolution in 19th century mathematics, which took us from one unique Space to a multiplicity of 'spaces', and from one true theory wearing its intended interpretation on its sleeves to wide-scope axiom systems. According to Beth, this came with profound changes in general intellectual methodology, making traditional philosophical "*What is X*" questions largely redundant – an anti-essentialist view which still seems worth expounding today. Another example which still seems highly relevant to me is his analysis of the 'Problem of Locke-Berkeley', putting metaphysical talk of arbitrary triangles to rest in a clear understanding of the role of variable and quantifiers in mathematical deduction. And as a final example of entanglement, I would mention Beth's masterful analysis of the methods of analysis and synthesis in mathematical history since Antiquity, their connection with Kant's notoriously difficult distinction between 'analytic' and 'synthetic' truths, and eventually, their harmonic coexistence in Beth's *semantic tableaux* – still a ubiquitous method in the field, which search for proofs and counter-examples at the same time. Beth's paper developing the latter has a lot of history (he likens them to the Tree of Porphyry) which modern referees would cut out at once as non-mathematical ballast and suspicious erudition. But for him, doing mathematical logic and doing philosophy was a natural unity. And this unity also worked the other way around: he did his mathematics driven by philosophical 'taste', rather than displays of cleverness per se. His famous Definability Theorem is an answer to a real issue about a whole field: how logic can be a systematic theory of *definition* as well as *deduction*. And likewise, his 'Beth models' for intuitionistic logic, the first plausible semantics for what Brouwer meant, high-lighted the informational processes behind constructive mathematics, a philosophical insight as much as a mathematical one. People with Beth's intellectual span still exist, though seeking mathematical respectability as such may have become more of a dominant norm.

Now to my own answers! How one thinks that mathematics influences, or should influence philosophy, depends first and foremost on what one considers essential *to mathematics itself*. And in this respect, I have my doubts about (or let us say, a lack of affinity with) many of the themes usually taught in courses in the philosophy of mathematics. Is the nature of *mathematical objects* important: are they Platonic entities, free constructions of the mind, nominalist fictions? I think it matters not at all to mathematical practice, and what is worse, concentrating on 'What is X' questions like this misrepresents what mathematics really is, and what makes it so appealing. Likewise, is *mathematical knowledge* a web of formal theories, connected by various embeddings and translations? Again, I think this view, despite its endorsement by non-logicians like Bourbaki, is a highly loaded description coming out of foundational research, rather than an unbiased account of what mathematics consists of. In particular, it cuts the field into disjoint territories like 'algebra', geometry', and so on, which often do not represent natural boundaries. Indeed, both earlier views put the cart before the horse. Mathematics is first and foremost a *human activity*, a family of ways of thinking (geometrical, combinatorial, and so on), a way of phrasing things abstractly, and a set of methods, aided by a store of accumulated results. Of course, this activity has given us a constant stream of intellectual *products*: definitions, theorems, theories, and so on – but the heartbeat of the field is in those activities, and in the creative patterns which they exhibit. The know-how is inextricable from the knowledge that. In the light of all this, my own analysis of how mathematics can or should influence philosophy would start from the mathematician's modus operandi, and then look at its repercussions in philosophical methodology, because. Moreover, – and any reader worth her salt must have seen this coming – I would make the same unconventional *"Activities over Products"* claim about philosophy in general. It is about intellectual activities: styles of analysis, defining, and reasoning, rather than any warehouse of special propositional insights.

Now to the Great Themes which I see as crucial to the flow of ideas from mathematics to philosophy. I will just give a few examples. Beth emphasized the philosophical, and indeed the general cultural importance of the 19th century move toward pluralism and creation of abstracts models for given postulates: spaces, algebras, groups. I would emphasize one that I think even more important, also from Beth's own interest in the 'definability' aspect of logic – namely, the theme of *structural invariance and the genesis of language*. When Helmholtz made his great proposal that geometrical language arises because of existing invariances in space, he set into motion what is arguably one of the most powerful mathematical perspectives. As set forth in Klein's "Erlanger Programm" (much more influential than "Hilbert's Program") any mathematical theory arises by defining a class of structures plus an equivalence relation over these, often given via a group of transformations. And the 'natural properties' of structures which arise in that way are the invariants of these transformations, which then give rise to languages defining these at the right level of expressive power. In particular, one mathematical structure, say Euclidean Space, can be analyzed at various grain levels in this way, from finer (as in geometry) to coarser (as in topology). This I think is absolutely crucial, and indeed, it fits some philosophers' ontological interest in a careful analysis of 'criteria of identity'. The invariance-to-language perspective still shows its power all the time, and across many disciplines: from cognitive psychology to image processing, and from new parts of mathematics to process theories in computer science. Defining abstract structures is just half the job, studying their interrelations at the right level, and then looking at the matching languages is the other.

A special case of this perspective has made it into philosophy over the last decade, viz. the study of 'logical constants' as the invariants under the broadest class of structural transformations, say all permutations, or even all homomorphisms from objects to objects. This powerful idea has come up independently in computer science, formal semantics, and the philosophy of logic, and, though I myself find the invariance criterion necessary but clearly insufficient, I think this mathematical style of thinking has lifted philosophical discussions about logical constants to a much higher level. Even so, I must admit to a sense of bewilderment when my colleagues in this area then go on to use these wonderful definitions in their quest for an Eldorado: the location of the 'exact border-line between Logic and Mathematics': a non-issue if I ever saw one. Who cares, as long as there is lively intellectual smuggling and traffic?

Another grand theme in mathematics which I see as crucial to philosophy is, of course, the notion of deductive argument and *proof* as a way of organizing compelling thought. I will not pursue this in detail, except to note that thinking about the structure and variety of exact inference has been a constant source of philosophical inspiration. Bolzano's "Wissenschaftslehre" (again a 19th century achievement, I really love that period!) comes to mind as a masterful account of the variety of intellectual reasoning styles, acknowledged and expanded in the work of Peirce, and finding its way eventually into the modern study of varieties of common sense reasoning by McCarthy and others in AI. In the latter area, it was clearly mathematical notions and results, rather than 'natural language philosophy' or 'common sense talk', which has made for genuine new insights. – "But" you might object, "this is AI, not philosophy!" Well, the way I see it, Von Clausewitz is right, here as ever: "AI is philosophy continued by other means". – In the same stream of mathematical modeling for inference, I see the influence of mathematical proof-theoretic paradigms, classical, constructive, linear logic-based, or more general sub-structural, on theories of meaning, knowledge, and nowadays even interaction – but other colleagues in this Volume will no doubt make this case much more eloquently.

Still, it seems fair to say that there is much more serious structure to mathematical proof than has been brought to light in the areas I have cited, including strategies for definition, ways of revising arguments in response to counter-examples, and finding pivotal structures for generalization. This has been pointed out by Lakatos and many other free-thinkers at some distance from modern logic, and these aspects too, seem worthy of serious philosophical reflection. I have formulated some thoughts on this in my 2006 *Topoi* paper "Where is Logic Going, and Should It?", which I will not repeat here.

Next, what about meta-mathematics? Is not that the greatest mathematical influence on philosophy of all? Here too, I will defer to other authors in this volume. Let me just say this on my own behalf. The great body of meta-mathematical results about the logical structure of mathematical and scientific theories, the families they form and their inter-theory relations, seems highly relevant far beyond the foundations of mathematics. It pertains to any serious philosophical discussion of the structure of knowledge, and issues like reduction, explanation, or other key themes in the philosophy of science. Perhaps not surprisingly, Beth's work, like that of Carnap, Peirce, or Bolzano, was an inextricable mixture of logic and philosophy of science. And let's not forget, even Tarski's first textbook was called "Logic and the Methodology of the Deductive Sciences". I feel this role of pure logic has been neglected, with philosophers of science going their own way – leaving logicians just the philosophy of mathematics as their private preserve.

But there is much more than standard meta-mathematics. From my own work on modal logic (clearly a branch of mathematics in its theoretical aspects), I would mention the role of epistemic, dynamic and temporal logics in epistemology, belief revision theory, or the philosophy of action, and a whole range of philosophical topics besides. Moreover, these formal ideas link up naturally with influences on philosophy coming from other areas of mathematics, such as learning theory (witness Kevin Kelly's work) or Information Theory (witness the work of Fred Dretske). Of course, I am not saying that every mathematical issue or method means something deep philosophically. But is seems good strategy to see if it does. For instance, in contemporary modal logic, one of the most delicate issues is the following additional dimension to the earlier-mentioned language – structural invariance duality. There is also the *Golden Rule of Balance*: the greater the expressive power of a logical language (tied to a finer structural equivalence), the greater its *computational complexity*. And once the language can say a whole lot of things, its logic will have become undecidable. This trade-off between expressive power and complexity is a deep phenomenon, whose true laws are yet poorly understood. What it adds to the research agenda is awareness of the *algorithmic* aspects of any semantic structures and informational tasks we choose to study. While I see this further dimension as absolutely central, algorithmics and complexity theory have failed to make any significant impact in philosophy so far that I can see.

Let me conclude with something closer to my home turf. Based on the influences that *have* occurred over the last decades, I would personally make the case that what is called 'philosophical logic' has been largely an interface area developing a mixture of mathematical methodology and philosophical sensitivity: a 'buffer state', if you wish, between mathematics and philosophy. I have chronicled the history of some relevant themes at the interface of Logic and Philosophy over the last century in my article in the *Handbook of the Philosophy of Logic*. There is a surprising amount of these, far beyond what one finds reflected in standard texts in philosophy of logic or mathematics, and together they show that one can have surprising, sometimes tortured, but often highly productive relationships here.

Of course, my list of influences is slanted toward logic – and there are many further areas of contemporary mathematics with proven philosophical impact, such as Probability Theory, Decision Theory, Game Theory, or Information Theory. Hence, I would be in favour of philosophy curricula offering a broad non-parochial band-width of formal methods. Philosophers should not just be able to appreciate logical deduction, but also probabilistic, game-theoretic, complexity-theoretic, and other mathematical reasoning. But who am I to make this plea? Logic evidently has no monopoly on the mathematics-philosophy interface, but let the other mathematicians speak for themselves.

What is the proper role of philosophy of mathematics in relation to logic, foundations of mathematics, the traditional core areas of mathematics, and science?

As I said already, I find it hard to keep some of these categories apart. Sometimes, I even fear that courses in Philosophy of Mathematics are just courses in mathematical logic made palatable to philosophy students by drawing all of its teeth. In any case, what I do find improper is the selling of in-house features of mathematical systems as philosophical issues. For instance, I am always amazed by the way some logicians have managed to get philosophers excited about issues like 'existence as the range of quantification', about 'meaning as what you get in a natural deduction proof', or about the 'border-line (another boundary, it's really an obsession!) between first-order and higher-order logic' – as if these were philosophical issues, rather than mainly internal technical system concerns for logicians, of no great outside significance that I can see – neither to mathematicians nor to philosophers.

What do you consider the most neglected topics and/or contributions in late 20th century philosophy of mathematics?

Here is what I miss most, in the form of three laments.

First of all, some sustained reflection on the *received agenda in the foundations of mathematics*. The way the history has gone 'After the Fall' by Gödel's Theorems has been an impressive sequence of insights into formalized theories, and the development of rich sub-disciplines of Recursion Theory, Model Theory, and Proof Theory. Sure, all this is truly admirable – and a more culturally rewarding 'catastrophe' than the demise of Hilbert's Program is hard to imagine. But still, there are also other roads that could have been taken, and that are still there for us to walk.

For instance, looking back at the foundational era, I have always been amazed by the almost pathological fears of inconsistency. Frege says somewhere that, if a single contradiction were to be discovered in mathematics, "the whole building would collapse like a House of Cards". Please, why? This claim seems largely an artefact of the wrong metaphor. Mathematics is not a house with foundations which have to bear the whole weight. It is rather a *planetary system* of different theories entering into various relationships, and happily spinning together in logical space. Damage one, and the system will continue, maybe with some debris orbiting here and there. Or, here is another metaphor for mathematics, equally attractive, due to Chaim Perelman: it is a wonderful *tapestry* of many strands woven together by the great mathematicians. Pull out one strand, and the tapestry may be weaker by an epsilon, but tears can be mended. And this brings me to my most central objection: we know from the history of mathematics and the sciences that contradictions are never the end of a story. To the contrary, one of the most striking ability of scientists is not to create infallible theories, but rather, having creative ways of coping with problems once they arise. In a wonderful, little-known study in the early 1960s, the Czech philosopher of law Ota Weinberger pointed out the persistent strategies for removing inconsistencies that can be found both in common sense reasoning (removing disagreements in conversation) and in science. Most of them go back to medieval logic and beyond. Trivially, one can give up some assumptions, the way Zermelo-Fraenkel Set Theory gave up Cantor's Full Comprehension for the Separation Axiom. Other strategies include making distinctions between kinds of objects that had been identified before, such as 'sets' versus 'classes' in NBG Set Theory. Another powerful strategy is the introduction of 'hidden variables', such as contextual arguments: I am tall for a human, but not tall for an animal. The history of science is replete with these, and much more inventive strategies. Where is the problem of foundations then, which we teach our students to gasp at in religious awe? I say that, on the basis of all this experience, that, if an inconsistency were to be discovered in, say, Peano Arithmetic this very evening, as I am writing this line, a marvelous analysis would be made within a short time, and an incomparably more subtle New Arithmetic would be found. Indeed, I often worry that *not enough* paradoxes and inconsistencies are being discovered these days to keep our most innovative mathematicians on edge ... . In this light, logic should also have studied patterns of repair, and ways of changing threatened theories. But more on that below.

A second conspicuous lack is the closedness of the agenda in other directions. To name the most spectacular of these, to me, one of the most spectacular new areas of mathematics in the 20th century has been the emergence of modern computer science (or *informatics*, as we say in Europe), with its fast-growing set of mathematical insights into computation, algorithmics, complexity theory, data structures, process theories, and so on. Of course, philosophers do discuss insights from recursion theory, and the like – but the fact that computer science has transformed our whole fundamental thinking about information, computation, 'information dynamics' (the term used by Milner and other protagonists), and highly sophisticated general theories of sequential and parallel action, far beyond the era of Turing and Church, seems to have bypassed the philosophy of mathematics – and general philosophy – altogether. The only claim one sometimes hears is that this is really about 'constructive mathematics', accompanied by wholly false claims that modern computer science is deeply dominated by intuitionistic logic, type theory and the like. Nothing like that is true! Indeed, the reverse should happen: insights from the study of information and computation are already informing logic, and they should also come to inform philosophy. This point is now slowly emerging in current attractive programs for a 'Philosophy of Information', on which a separate *Five Questions* volume is being published.

This closed nature of the agenda is also being reinforced by the current wave of historical research into the Golden Age of the foundations of mathematics. While this may seem a harmless pursuit aimed at increasing our understanding how we got to be here, I also sense another, much more problematic undercurrent in this whole historicizing trend. 'The only issues of real value are those from the past, and the variations we can find for them'. In this nostalgic way, Frege, Gödel and Turing are still made into our teachers, rather than the innovative thinkers of today changing the agenda of logic.

Finally, now that I am in complaining mode, let me step on the accelerator full throttle. The philosophy and foundations of mathematics, it seems to me, show a curiously defensive attitude, ill-fitting their status and achievements. Friendly proposals to extend the classical agenda (we cannot keep singing hymns to Gödel and Tarski *forever* – unless we are already in Heaven) are perceived as threats, to be received with suspicion and sometimes even personal attacks. And innovative 'agents provocateurs' like Lakatos in his mind-opener "Proofs and Refutations" were subjected to torrents of abuse, rather than the lively interest in new ideas one would expect from a vigorous community. As Samson Abramsky once put it, the established order in foundations reminds him of what was said about the Bourbon Dynasty, when they re-entered France again after the Revolution: "They had learnt nothing, – and they had forgotten nothing". Indeed! I myself am often struck by the analogies between modern attitudes in foundational research and Islamic fundamentalism. Both insist on a mythical purity of the past, sanitizing our founding fathers of their (Heaven knows) many ambiguities and weaknesses, and both claim that if we can only stick to the past, ignoring the key ideas transforming the modern world outside, all evils will be cured. To be sure, I see myself as a logician, and a great lover of mathematical technique, but I have never been able to understand this defense of the status quo. In particular, I have always found many outspoken critics of modern logic (Blanshard, Perelman, Toulmin, Lakatos) extremely interesting and well-worth reading, and a useful reminder of the many doors our Founding Fathers have closed historically – doors that could be opened again now, *precisely because* we can feel confident in what has already been achieved.

What are the most important open problems in the philosophy of mathematics and what are the prospects for progress?

Here is what I take to be the main challenges for the future in the philosophy of mathematics, and indeed foundational research. I think we are still far from a true understanding of mathematics and what it can mean for philosophy. I have to formulate the following points somewhat apodictically, for lack of space – but

I have some publications backing them up, for instance, 'Logic and Reasoning: Do the Facts Matter?' in a forthcoming issue of *Studia Logica* edited by Hannes Leitgeb, and 'Intelligent Interaction: Logic as a Theory of Rational Agency', to appear in the *Proceedings of the 2007 DLMPS Conference* held in Beijing (an organization founded by Beth and Tarski, by the way).

First, mathematics is not some isolated faculty of the human mind which needs to be approached with special reverence. It is rather an *in vitro* version of one crucial faculty of rational agents in general, and I think we need to embed our understanding of mathematics in more encompassing theories of rational agency. Thus, philosophy of mathematics and epistemology should go hand in hand, but both with suitably broad agendas. I myself see a fluid transition all the way from common sense reasoning to mathematical proof, and from knowledge structures in daily life to mathematical theories. Accordingly, we should arrive at an integrated understanding of that whole spectrum. Moreover, but I guess this point is too obvious to need much elaboration, we should not take deduction as our only model here, since intelligent handling of information also involves precise observation and communication.

Next, in line with what I said about 'foundationalism', the true measure of intelligent human behaviour, in my view, is not to be always right, and to have infallible procedures keeping us on that track forever. We see human intelligence at its finest when we *correct ourselves*, learn from mistakes, and create something new and better out of broken dreams and refuted expectations. These processes are studied nowadays under the heading of belief revision and similar parts of logic, but they apply equally well to a real understanding of mathematical practice. In medical terms, the foundationalists wanted to make the whole world of mathematics free from disease, by killing off every last unclarity and inconsistency. I say, in contrast, that we should study the ubiquitous 'sanitary' processes of mathematical *precisation* and *revision*, which swing into action every time a new problem is discovered. Mathematics is also about processes of correction, invention of language, precisation when clarity is to be saved, but also less formal paraphrase when an overview is needed of a proof or a notion. Accordingly, in my favourite slogan, logic is not just the guardian of being right, its true role is much better described as follows: *logic is the immune system of the mind!*

My third agenda item is this. Mathematics is a social activity, just as much as any other branch of science. The true locus of mathematical progress is the *seminar room*, just as much as the attic with a Rodinesque thinker. Thus, we are talking multi-agent interaction, including human subjects, and not just one, but many of them. Indeed, this theme has been hidden under the floor boards ever since the Golden Thirties. Carnap stressed the 'inter-subjective nature' of scientific insight, but how to explain that when our logical theories have no explicit subjects accounted for? Likewise, even Turing, when discussing his famous Test whether a machine could mimic a human, emphasized the essentially social nature of learning and scientific progress. I expect that a true understanding of mathematics must be embedded eventually, not just in a theory of self-correcting rational agents, but in a theory of intelligent interaction.

My fourth point is the role of the empirical facts. As we are rapidly learning more about the actual functioning of mathematical abilities in the human brain, *cognitive science* enters the picture. We are fast learning about the delicate interplay of visual, linguistic, and planning components of the brain in meaningful reasoning activities, as well as counting and measuring. I think a modern philosophy of mathematics must deal with this new source of information, which replaces centuries of speculation about intuition and the like, in just the same way as other areas of philosophy, and indeed, logic itself.

But let me end with something simple, and almost pedestrian. If we are going to do philosophy of mathematics today, it might be good to *just take a simple look at what mathematics today really is*, in the hands of its best practitioners – and decide on its major features only afterwards. Some years ago, I decided to teach a public evening course on modern mathematics with my distinguished Amsterdam colleague Robbert Dijkgraaf, our leading expert in string theory and our best-known mathematician nationally. The reason was that we had had a conversation in which we both felt that the usual division of the field into areas like Algebra, Geometry, etc. gives no true picture of the intellectual geography of the field. We both thought the essence lay elsewhere, in ubiquitous notions and patterns. We decided on an experiment. We both drew up a list of what we considered key aspect of mathematics, and if those lists would overlap enough, we had a basis for a joint course. Indeed, the two lists, one by a mathematical logician and one by a mathematical physicists, were strikingly similar. And here are the topics that we taught: (a) Symmetry, Invariance, and Language, (b) Counting with Numbers and with Language, (c) Order,

(d) Proof, (e) Computation and Complexity, (f) Paradoxes and Meta-theorems, (g) Probability, (h) Games and Information Dynamics, and (i) Prediction and Dynamical Systems. Each evening, we would show how these topics arose in Nature (with an emphasis on physics), and then in Cognition (with an emphasis on logic), and the third hour would be for discussion with our audience. I have seldom felt the vibrancy and cultural impact of mathematics like in this evening course. I wish the Philosophy of Mathematics could achieve equal vibrancy, doing justice to the true attractions of its subject.

5

Douglas Bridges

Professor of Pure Mathematics

University of Canterbury, New Zealand

Why were you initially drawn to the foundations of mathematics and/or the philosophy of mathematics?

I was probably interested from the time when, as a fourteen-year-old who had just discovered that he could do mathematics and that he found it fascinating, I read W.W. Sawyer's classic "Prelude to Mathematics" [3].[1] There the author discussed the nature of mathematics and, if I recall aright, came up with the view (one I still find attractive) that mathematics is essentially the study of patterns—geometric, algebraic, or whatever.

My interest in foundations really took off when I bought Errett Bishop's "Foundations of Constructive Mathematics", on the only occasion that I saw it on sale, in James Thin's bookshop in my home town of Edinburgh. At that time I was in my first year as a Ph.D. student working on von Neumann algebras, and had been experiencing an unarticulated sense of dissatisfaction with a kind of proof that was common in that subject: one where you use Zorn's lemma to construct a maximal family of objects, typically projections, with a certain property, often having first supposed the contrary to what you want to prove. When I read

---

[1] In an online interview, the logician and mathematical expositor Keith Devlin writes: "... W. W. Sawyer's Prelude to Mathematics (Penguin, 1955) ... together with another Sawyer book, Mathematician's Delight (Penguin, 1943), was pivotal in making me decide to become a mathematician. Although the author's 1950s writing style comes across as stilted today ..., for me, reading them as a teenager in the early 1960s, every page sang as Sawyer brought advanced mathematics to life". These words fit my case exactly. For Devlin's interview go to
http://www.americanscientist.org/template/ScientistNightstandTypeDetail/assetid/44767;jsessionid=aaa5LVF0.

Errett's preface and first chapter, provocatively entitled "A Constructivist Manifesto", I understood, and became able to articulate, my dissatisfaction: the aforementioned proofs, subtle and cunning though they were, by their very nature did not provide any information about the construction of the objects whose existence they "proved". My initial encounter with Bishop's work led me eventually, after I had abandoned von Neumann algebras for a few years of high-school teaching, to a D.Phil. in constructive set theory and analysis, and thence to thirty-plus years of research in constructive (and, occasionally, classical—I'm not a fundamentalist!) mathematics.

What examples from your work (or the work of others) illustrate the use of mathematics for philosophy?

In the case of constructivism, which came first: the mathematics or the philosophy? It could be argued that once mathematicians had recognised the distinction between constructive and nonconstructive arguments (a recognition that certainly predates the major players in the constructive-versus-nonconstructive philosophical debates), they were on course to examining the philosophies that might underlie that distinction. It could also be argued that mathematicians like Kronecker and Brouwer were led to constructive mathematics by their ontological and consequent epistemological beliefs.

I guess my and others' work in constructive analysis, algebra, and topology has indirectly fostered philosophical debate, in that, without a body of constructively developed mathematics, it would be hard for a philosopher to generate interest in discussion of the issues surrounding constructivism. Whether our work has had any more direct impact on the philosophy of mathematics, I leave to the philosophers to judge.

What is the proper role of philosophy of mathematics in relation to logic, foundations of mathematics, the traditional core areas of mathematics, and science?

It's not easy to answer that, as it depends on what one means by "proper". I like to think that future generations of mathematics will pay much more attention to foundational/philosophical issues, at least to the extent that, from time to time, they examine the meaning of their work—and by "meaning" here I do not mean (!)

simply the interpretation of their results within mathematics or some subject to which they can be applied: I mean the intrinsic meaning of their work.

For example, suppose I prove a theorem stating that the range of a holomorphic function f with an isolated essential singularity at the centre of a disc D in the complex plane $\mathbb{C}$ is either the entire plane or else the plane minus a single point. Can my proof tell me, for a given f, which of the two alternative conclusions holds? If not, what is the real meaning of my theorem? In fact, no proof of Picard's theorem—which is what we're talking about—can provide a procedure for deciding which conclusion holds, so the theorem, as normally proved in the literature, can be regarded as saying that it is impossible for the range of f to exclude more than one point of $\mathbb{C}$. But there *is* a proof that embodies an algorithm with the following property: given two distinct complex numbers ξ and ξ', the algorithm will output a number z in the punctured disc D and will show, *by computation*, that either $f(z) = \xi$ or $f(z) = \xi'$; see [2]. This property is logically equivalent to the standard conclusion of Picard's theorem, but embodies significant computational information that the standard proofs of that theorem cannot provide.

If pure mathematicians continue to carry out research without paying any attention to issues of meaning and related ones of ontology, then, however technically brilliant and exciting (to mathematicians) their work may be, they run the risk of their subject becoming like Hermann Hesse's *Glassperlenspiel*, an esoteric game with no connection with physical or mental reality (whatever they may be!). The pure mathematician will have "followed the gleam and it [will have] led him out of this world" ([1], page viii).

What do you consider the most neglected topics and/or contributions in late 20th century philosophy of mathematics?

I must be careful to emphasise that what I say here is not a grumble about any lack of recognition, perceived or otherwise, of my own research.

I believe that, in spite of the increasing interest in constructive issues, theoretical and practical, among logicians and computer scientists, working pure mathematicians remain, for the most part, uninterested in the meaningful distinction between constructive and nonconstructive interpretations of "existence". There is, in my experience, a widespread opinion that, while constructive proofs are all very nice to have in principle, in practice they cannot be carried out in the technically more demanding parts of modern mathematics. This opinion, which may well be a continuing reflection of Hilbert's prestige and influence, is based on a neglect of the modern mathematical/philosophical/foundational literature dealing with constructive issues. In extreme cases—fortunately, a lot rarer these days than was the case thirty years ago—it leads to a dismissal of constructive issues as irrelevant to "real" mathematics. This is highly regrettable and simply wrong, for at least two reasons. First, Bishop's book alone showed that constructive methods can be used systematically to develop deep mathematics across a broad spectrum of areas. Secondly, every constructive proof embodies potentially significant information that perforce cannot be provided by a nonconstructive one; indeed, one can extract from a constructive proof an implementable algorithm, with the bonus that the proof itself verifies that the algorithm meets its specifications.

I'm not for a moment suggesting that all mathematics should be done constructively (as I said above, I'm not a fundamentalist). I *am* advocating that all pure mathematicians should, first, admit that there is a meaningful distinction between constructive and nonconstructive existence, and, secondly, accept that it is a worthwhile mathematical activity to examine the constructive content of mathematics.

No doubt there are other neglected topics/contributions in late 20th century philosophy of mathematics; but I do not feel competent to identify or discuss those.

What are the most important open problems in the philosophy of mathematics and what are the prospects for progress?

The most important questions that spring to my mind are these.

- **What are the distinctive features of a mathematical proof?** Is a computer-based argument like the one used for the four-colour theorem really a proof? Are theorem-proving expert systems acceptable as a means of mathematical proof? Can we even *define*, precisely, what is a proof?

- **What are mathematical objects?** We have the continuing debate over Platonism, formalism, intuitionism, etc, etc;

but we still seem no further along the path to any consensus about what, say, the number "one" is.

- **What are the distinctive features of mathematics itself?** How do we distinguish mathematics *per se* from disciplines that *use* mathematics? I'm thinking here about "modelling" for example, which is currently a big deal in some mathematics departments. Is modelling such things as blood flow, fish population dynamics, or the growth of cancer tissue "mathematics", or is it the application of mathematics in other disciplines? My feeling is that it is the latter;[2] but what, if any, justification is there for such a feeling? Justification could only arise out of a clear definition of what characterises mathematics as a discipline.

To conclude: it will be obvious to philosophers that I am not one of their number, much as I enjoy thinking about the issues raised above. They should be patient with a mere mathematician who has a strong interest in those issues, even if he lacks the philosophical competence to handle them rigorously!

5.1 REFERENCES

[1] E.A. Bishop, *Foundations of Constructive Analysis*, McGraw-Hill, New York, 1967.

[2] D.S. Bridges, A. Calder, W. Julian, R. Mines, and F. Richman, 'Picard's theorem', Trans. Amer. Math. Soc. **269**(2), 513–520, 1982.

[3] W.W. Sawyer, *Prelude to Mathematics*, Penguin Books, 1955. Reprinted by Dover Publications, 1982.

---

[2] As I mentioned earlier, some people say that, for example, my type of work in constructive mathematics is not "real" mathematics—a real put-down if ever there was one. I'm actually saying here that modelling is *not mathematics at all*! But this is not a put-down, as I would not for one moment suggest that modelling is not an important mathematical activity. The question for me is whether modelling is "mathematics" or "mathematical".

6

Charles S. Chihara

Emeritus Professor of Philosophy
University of California at Berkeley, USA

What is the proper role of philosophy of mathematics in relation to logic, foundations of mathematics, the traditional core areas of mathematics, and science?

To answer that question properly, I feel that I first need to present briefly what I take philosophy of mathematics to be.

The philosopher seeks an understanding of the world. But the sort of understanding sought, I like to call "Big Picture understanding". The philosopher seeks a kind of general view of the world that results from answering such questions as: What, in general and in broad outlines, is the universe like? What, in general and in broad outlines, is our place in the universe and how are we related to the things around us? How, in general and in broad outlines, are we humans able to gain an understanding of the things around us and of the universe? Of course, there are also many areas of philosophy that are researched by specialists. So there is philosophy of science, of mathematics, of psychology, of language, of ... Someone working in the philosophy of X will seek a Big Picture understanding of the nature of X—one that will aid philosophers in achieving the goal of fitting X into the Big Picture of the relationship between humans and the universe they inhabit. Thus, the philosopher seeks ultimately to obtain an account of X that fits together with the accounts of the other Ys which she accepts, yielding a consistent and coherent account that is compatible also with the epistemology and natural sciences that she judges to be acceptable.

Consider, then, the specific area of philosophy that concerns us here. The philosopher of mathematics wants to gain a Big Picture understanding of the nature of mathematics—one that will result in a conception and theory of mathematics which will be compatible with not only what science teaches us about how we humans acquire our knowledge and understanding of the world around us, but also with our views about, and knowledge of, science itself.[1]

Thus, the traditional core areas of mathematics, including number theory and analysis, can be seen to be target areas for analysis by the philosopher of mathematics as she attempts to develop an adequate Big Picture account of mathematics. What about logic (that is, mathematical logic)? Logic can be regarded as providing an excellent example of how mathematics is used to model important aspects of human intellectual life—in this case, "deductive thought".[2] Logic can thus be seen to be a target area of analysis for the philosopher, since mathematical modeling is an important and widely employed use of mathematics. Since logic can also be regarded as "the study of the type of reasoning done by mathematicians" ((Shoenfield, 1967), p. 1), logic should be, like science, an area with which the philosopher's account of mathematics should cohere.

Logic also is one of the chief topics studied in the area of mathematics known as *foundations of mathematics*, where specific theorems of logic (such as Gödel's Incompleteness Theorems) are discussed; as such, logic is also an important part of a target area of the philosopher's investigation.

But what is "foundations of mathematics"? Courses and books in the foundations of mathematics cover topics that are also covered in philosophy of mathematics courses and books. What then distinguishes foundations of mathematics from philosophy of mathematics? Foundations of mathematics courses are generally taught by mathematicians, whereas philosophy of mathematics courses are generally taught by philosophers. Books whose titles contain the phrase 'foundations of mathematics' tend to have much more mathematical content than do books whose titles contain the phrase 'philosophy of mathematics'; whereas books whose titles contain the phrase 'philosophy of mathematics' tend to be much more philosophical in nature than do books whose titles contain the phrase 'foundations of mathematics'.[3] In view of these differ-

---

[1] The above is an all too brief account of my view of philosophy. The industrious reader can find a fuller and more adequate account in the Introduction to my (Chihara, 2004).

[2] See (Enderton, 1972), p. 1, for an account of this view of logic.

[3] Here are a just a few works that are clearly foundations of mathematics books: .(Wilder, 1952), (Beth, 1959), (Hatcher, 1968), (Bar-Hillel, Posnanski, Rabin, & Robinson, 1966), and .(Mancosu, 1998). Kneebone's book (Kneebone, 1963) is based upon a series of lectures attended by both undergraduate

ences, how should one characterize foundations of mathematics? The mathematician Stephen G. Simpson suggests that:

> foundations of mathematics" means the study of basic mathematical concepts (quantity, number, geometrical figure, etc), how to organize these concepts into a hierarchy of more and less fundamental concepts, how to set up axioms and rules of proof for mathematics, and in the systematic phase, a study of the properties and limitations of such formal systems.[4]

Simpson's conception of foundations of mathematics fits nicely with some aspects of what is done in some classic books on the "foundations of mathematics", but it does not seem to jibe with other features of such books. For example, the first part of Raymond Wilder's book *Introduction to the Foundations of Mathematics* roughly fits Simpson's description, but the second part seems to be more like a philosophy of mathematics book. Interestingly, Wilder seems to regard the second part as covering "Foundations proper" ((Wilder, 1952), p. ix), in accord with the following characterization of the field: "Scientific inquiry into the nature of mathematical theories and the scope of mathematical methods"[5]

If one regards foundations of mathematics in the way Simpson does, then the field can be regarded as an aid to the philosopher of mathematics by providing analyses of fundamental concepts of mathematics. Also, being an area of mathematics, foundations can be regarded as, itself, a target area of philosophical analysis. On the other hand, when foundations is understood in the way Wilder favors, the area can be regarded as a version of philosophy of mathematics, forming part of the ongoing program of investigating the nature of mathematics, but undertaken from the perspective of the mathematician.

---

and graduate students; it is divided into three parts, the second of which is entitled "Foundations of Mathematics".

[4] This quotation is from Stephen G. Simpson's "What is Foundations of Mathematics?" at the web site (dated 7 April 2006):
http://www.math.psu.edu/simpson/hierarchy.html
retrieved: November 1, 2006.

[5] "Foundations of mathematics" (2006). In *Encyclopedia Britannica*. Retrieved November 2, 2006, from Britannica Concise Encyclopedia: http://concise.britannica.com/ebc/article-9371531

What examples from your work (or the work of others) illustrate the use of mathematics for philosophy?

I shall provide two examples of uses that I have made of mathematics in my own investigations. The first example is my use of a specific system of set theory, Hao Wang's Sigma omega,[6] in order to respond to what I call "Quine's challenge to the nominalist"[7]. Quine had argued in *Word and Object* that the nominalist would be unable to accept a scientific theory that required mathematics in its statement and application, unless the mathematics in question amounted to only trivial portions of arithmetic—this because mathematics (except for trivial portions of arithmetic) is "irredeemably" committed to the existence of abstract mathematical objects. Basically, my response consisted in putting forward a modal interpretation of Wang's system (by way of the notion of "constructability quantifiers"—a primitive of the interpretation) and in then showing how the axioms of the system under this interpretation could be seen to be true. More specifically, the sentences of Wang's mathematical system were systematically interpreted to yield statements about the constructability of open-sentences. It was then argued that the translations of the axioms of the system were true statements and that the rules of inference of the system preserve truth. It was then shown how a nominalist who accepted modal notions could make use of such a system of mathematics in her scientific theorizing.[8] Such a predicative system of mathematics would furnish the nominalist with the mathematical tools needed to develop a version of science in which most, if not all, of the reasoning of contemporary science could be carried out.[9] The nominalist could, in this way, use a much stronger system of mathematics than the trivial portions of arithmetic that Quine had allowed.

Let us now consider a different use of mathematics in the ser-

---

[6] One of a family of formalized predicative set theories described in Chapter 24 of (Wang, 1962). Wang's system Sigma can be regarded as a kind of union of the family members referred to above. More accurately, Sigma is a recipe for making formal systems of predicative set theories whenever an ordinal is given.

[7] See (Chihara, 2004), Chapter 5, Section 2.

[8] See (Chihara, 1973), especially Chapter 5 and the Appendix.

[9] There is growing evidence, due primarily to Saul Feferman, that much (if not all) of the mathematics of contemporary science can be reproduced within a predicative system. See, for example, (Feferman, 1998a), Sections 8 and 9, and (Feferman, 1998b). See also (Hellman, 1998), Sections 1 and 2.

vice of a nominalistic philosophy. In *A Subject With No Object*, John Burgess and Gideon Rosen analyze recent nominalistic work in the philosophy of mathematics. Their book focuses, especially, on the nominalistic reconstructions of mathematics and science that have appeared in the last twenty or so years—reconstructions such as my constructability version of Wang's Sigma omega (described above), Hartry Field's mathematical instrumentalist version of mathematics and physics (as developed in (Field, 1980)), and Geoffrey Hellman's modal structuralist version of mathematics (presented in (Hellman, 1989)). Burgess and Rosen take up the question of the value of these nominalistic reconstructions in the last chapter of their book, asking specifically what this body of nominalistic works is good for?

Burgess and Rosen respond to the above question by considering two possible replies:

(A) The reconstruction provides us with a hermeneutical "analysis of the ordinary meaning of scientific language". This is the response of the "hermeneutical nominalist" ((Burgess & Rosen, 1997), p. 206–8).

and

(B) The reconstruction provides us with an alternative version of science, which is better than, and to be preferred to, our present day versions of science. This is the response of the "revolutionary nominalist" ((Burgess & Rosen, 1997), III.C.I).

After finding both possible replies to be unacceptable, they then go on to write:

> Since anti-nominalists reject all hermeneutic and revolutionary claims, from their viewpoint the various reconstruals are all distinct from and *inferior* to current theories. What is accomplished by producing a serious of such distinct and inferior theories? *No advancement of science proper, certainly, ...*" ((Burgess & Rosen, 1997), p. 238, italic mine).

My reaction to the above quotation was to ask myself: What justifies their claim that the nominalist's reconstruals are all *inferior* to current theories? And why are they so sure that these reconstructions of mathematics make no scientific advancements?

It was in response to the reasoning described above that I explored the use of *nonstandard analysis* as an instrument of rebuttal.[10] Under the inspiration of Skolem's construction of nonstandard models of arithmetic, Abraham Robinson constructed nonstandard models of analysis. These constructions were carried out by extending the real numbers to a set that includes infinitely small and infinitely large numbers. The infinitely small numbers were then used as "infinitesimals" in the definition of such concepts of the calculus as limit, differentials, continuity, and integrals, thus following in the tradition of such pioneers in analysis as Newton and Leibniz.

Robinson's nonstandard analysis provides the framework for teaching calculus in a completely new way. Textbooks of elementary calculus have been published and classes in calculus have been given using the nonstandard approach. Kathleen Sullivan produced a study of an experimental group of five calculus classes from four small private colleges and one large public high school, all learning calculus from the perspective of nonstandard analysis, with a control group of students of comparable ability,[11] who were taught the material in the standard way. She summarized the results of her study in the following way:

> [T]here does seem to be considerable evidence to support the thesis that this is indeed a viable alternate approach to teaching calculus. Any fears on the part of a would-be experimenter that students who learn calculus by way of infinitesimals will achieve less mastery of basic skills have surely been allayed. And it even appears highly probable that using the infinitesimal approach will make the calculus course a lot more fun both for the teachers and for students. ((Sullivan, 1976), p. 375).

Let us now ask the Burgess-Rosen question of value: "What is nonstandard analysis good for?" Evidently, there are two principal

---

[10] See "The Burgess-Rosen Critique of Nominalistic Reconstructions" (forthcoming in *Philosophia Mathematica*), where this use of nonstandard analysis as part of my rebuttal to the Burgess-Rosen reasoning is presented in much more detail.

[11] The students in the control group were all from the same five institutions mentioned above. The judgment that the students of the two groups were of comparable ability was confirmed by their SAT Math Ability Scores (see (Sullivan, 1976), p. 373).

answers that should be considered:

(A) Nonstandard analysis provides us with a hermeneutical "analysis of the ordinary meaning of scientific language". This is the response of the *hermeneutical nonstandard analyst*.

and

(B) It provides us with an alternative version of science, which is better than, and to be preferred to, our present day versions of science. This is the response of the *revolutionary nonstandard analyst*.

By reasoning in the way Burgess and Rosen did for the earlier cases regarding the nominalistic reconstructions, we can infer that neither of the above answers is at all acceptable. Should we then say, as Burgess and Rosen wrote in the case of nominalistic reconstructions?

> Since anti-nonstandard analysts reject all the hermeneutic and revolutionary claims of the nonstandard analysts, from their viewpoint, *nonstandard analysis is distinct from and inferior to standard analysis*. What is accomplished by producing a series of such distinct and inferior theories? *No advancement of science proper, certainly,* ...

Such reasoning, however, is refuted by the judgments of many mathematicians, who have attested to the value of nonstandard analysis in producing important mathematical discoveries – discoveries such as the Bernstein-Robinson theory of invariant subspaces of infinite dimensional linear spaces, which settled an outstanding open question of many years standing.[12] Mathematicians have also argued that nonstandard analysis has been usefully applied in such fields as quantum theory, thermodynamics, economics, and probability theory, shedding new light on our understanding of mathematical questions and leading to new and simpler proofs.[13] All of this shows the dubiousness of thinking that a new and alternate framework for mathematical reasoning should be judged *inferior* to standard ways of mathematical thinking on the basis of concluding that the framework provides us

---

[12] See (Davis, 2005), p. 1.
[13] See (Dauben, 1988), pp. 192-3; (Ostebee, Cambardella, & Dresden, 1976), p. 878; and p. x of the Forward to (Robinson, 1996).

neither with an adequate hermeneutical "analysis of the ordinary meaning of scientific language" nor with an alternative version of science, which is better than our present day versions of science. As the above example shows, such a basis supplies us with no good reason for concluding that the framework in question can make no advancement of science.

Why were you initially drawn to the foundations of mathematics and/or the philosophy of mathematics?

I started my undergraduate studies as a civil engineering student. After only a year in the program, I realized that a career in engineering was not for me. So I switched to physics. However, by my junior year, I became dissatisfied with physics and changed to mathematics. After completing the bachelor's degree, I entered a PhD-program in mathematics. It was during my first year as a graduate student that I attended a colloquium talk on the foundations of mathematics. The days following found me in the mathematics departmental lounge discussing the colloquium over coffee with both students and faculty. That was the first indication that my deepest interest in mathematics concerned questions in the foundations of mathematics. Later, I took a lecture course in the foundations of mathematics (in which Wilder's book was the text) and a reading course in which Gödel's Incompleteness theorems formed the core material. Such courses prompted me to change the direction of my studies to the philosophy of mathematics.

My university studies thus went through the following sequence of changes: engineering, physics, mathematics, philosophy. Clearly, each step was a change away from the practical and towards the more theoretical.

What do you consider the most neglected topics and/or contributions in late 20th century philosophy of mathematics?

I shall not attempt the sort of ranking of neglected topics in late 20th Century philosophy that would enable me to select the *most* neglected of these topics. What I shall advance and discuss instead are a few neglected topics that I regard as both having some importance and also being of some interest to philosophers of mathematics.

In the late 20th century, there were a number of attempts to resuscitate and defend one or other of the principal schools of thought in the foundations of mathematics that dominated early 20th century philosophy of mathematics. For an example, Michael Dummett wrote extensively on Intuitionism in a way that prompted a number of philosophers, especially analytic philosophers in England, to defend (and in some cases promote) versions of Intuitionism.[14] Dummett's defense of Intuitionism was strongly influenced by a distinctive view of language and philosophy promoted by Ludwig Wittgenstein.

Now consider the conception of philosophy of mathematics I described at the beginning of this essay: the philosopher of mathematics, according to this conception of philosophy, wants to gain a Big Picture understanding of the nature of mathematics—one that will result in a conception and theory of mathematics which will be compatible with not only what science teaches us about how we humans acquire our knowledge and understanding of the world around us, but also with our views about, and knowledge of, science itself. The mathematics being referred to in the above characterization is *actual mathematics*: it is the kind of mathematics that is taught at every university in the world. But the mathematics of Intuitionism defended by Dummett is not "actual mathematics" but Intuitionistic mathematics—a distinctively different kind of mathematics that is seriously used in very few (if any) science courses in any university in the world and that is far weaker than the classical mathematics that L.E.J. Brouwer and other Intuitionists so forcefully denigrated. So a neglected problem for neo-Intuitionism is to provide a Big Picture account, as described above, of actual mathematics. Since it is actual mathematics, and not Intuitionistic mathematics, that is applied by scientists and engineers, the problem is to explain how it is that this non-intuitionistic mathematics is applied with such enormous success by scientists and engineers the world over. How could these successful applications be produced even though the mathematics being used is false? unintelligible? meaningless? or ... (substitute what other term your favorite neo-Intuitionist may use to negatively characterize classical mathematics)? This is a problem that those advocating Intuitionism should, but, so far as I know, never

---

[14] See (Posy, 2005) for a recent survey of Intuitionism and Dummett's special philosophical contribution to the development of the neo-Intuitionist philosophy. The reader can find many relevant references in Posy's article.

do, address.

Consider now the attempts made near the end of the 20th and at the beginning of the 21st century to resuscitate Frege's version of Logicism.[15] Frege wanted to show that the truths of arithmetic are all "analytic" (in a sense to be discussed below) by providing proofs, from the "logical laws" and definitions of his system, of a sufficient number of theorems to convince his readers that all the truths of arithmetic are provable in this way. It was obvious that his attempt failed when his system was shown to be inconsistent. The resuscitators (the "neo-Fregeans") have attempted to salvage the spirit of Frege's philosophy by, among other things, revising his system of logical laws and definitions so as to yield a consistent system, while preserving the key theorems of the original system. The principal idea was to replace Axiom V of Frege's system with what is called Hume's Principle.[16] Much has been written, both pro and con, about this attempt to save Frege's Logicism, but due to limitations of space, I cannot give an assessment of the disputes here. Instead, I wish to focus on a neglected aspect of the controversy.

What are the "analytic truths"? They are the truths provable from general logical laws and definitions (Frege, 1959).[17] Since Frege took it as obvious that any analytic truth is *a priori*,[18] we can assume that the totality of general logical laws is a sub-totality of the general laws mentioned in his definition of *a priori* truths. It then follows that a general logical law neither needs nor admits of proof. But what makes a general law a general logical law? Frege distinguished three sources of knowledge: sense perception, the logical source of knowledge, and the geometrical and temporal sources of knowledge ((Frege, 1979), p. 267), the logical source of knowledge being that from which our knowledge of the logical

---

[15] See (Hale & Wright, 2005) for a recent survey of these attempts, with references to the principle works both pro and con these attempts.

[16] See (Hale & Wright, 2005), p. 167, for an extensive discussion of Hume's Principle.

[17] p. 4. Hale and Wright note (in .(Hale & Wright, 2005), p. 166) that Frege explained what it is for a proposition to be analytic in the way explained above, and then go on to claim that their version of neo-Fregeanism "holds that [Frege] was substantially right" in maintaining that the theorems of the theory of natural numbers are analytic in that sense. But later, they explain they do not hold that arithmetic is analytic in "Frege's own strict sense", but only in an extended sense of that term (p. 169).

[18] A proposition is *a priori* if it is provable "exclusively from general laws, which themselves neither need nor admit of proof" ((Frege, 1959), p. 4).

laws ultimately must come.¹⁹ Furthermore, a general logical law must have "the widest domain of all": it must apply to everything thinkable.²⁰

What did Frege hope to accomplish by showing that the truths of arithmetic are all analytic? Among other things, he hoped to show that our knowledge of arithmetical truths is ultimately based not upon sense perception or upon our geometrical or temporal intuitions, but only on the logical source of knowledge. Had Frege succeeded, he would have shown something very fundamental and important about the nature of arithmetic that would have refuted a basic Kantian doctrine.²¹

Now the neo-Fregean's belief that Frege was essentially correct in claiming that the truths of arithmetic are analytic is based on the fact that the sentence labeled 'Hume's Principle' (henceforth 'Hume's Sentence') adjoined to a suitable formulation of second-order logic yields a system in which all the "fundamental laws of arithmetic" – including Peano's axioms – are derivable.²² Let us call this result, basically following Wright, "Frege's Theorem". The question is: what is the status of Hume's Sentence? Does it express a general logical law? No, the neo-Fregeans make no such claim. Instead, they claim that it expresses a definition: an "im-

---

[19] One can see why Frege would write in the concluding section of (Frege, 1959), p. 99: "Arithmetic thus becomes simply a development of logic, and every proposition of arithmetic a law of logic ..." One can also see why Frege would think that his conception of analytic propositions was essentially the same as that of Kant's ((Frege, 1959), p. 3, fn).

[20] I infer this from Frege's claim that: The truths of arithmetic govern all that is numerable. This is the widest domain of all; for to it belongs not only the actual, not only the intuitable, but everything thinkable. Should not the laws of number, then, be connected very intimately with the laws of thought? ((Frege, 1959), p. 21). Of course, this view of logical laws is connected with Frege's idea that the laws of number are not really applicable to external things; they are not laws of nature. They are, however, applicable to judgements holding good of things in the external world; they are laws of the laws of nature. ((Frege, 1959), p. 99).

[21] Strictly speaking, Frege's reasoning would not have directly refuted Kant's claim about the synthetic nature of arithmetical truths (since Kant used a slightly different characterization of "analyticity" in making his claim), but it seems clear that he certainly would have refuted the spirit of Kant's claim. See (Frege, 1959), Part 5.

[22] The proof of this fact is due to Frege (see (Wright, 2001), p. 273). A second-order formulation of Hume's Principle is given there (see fn. 5). Informally, it is supposed to say:
For any concepts F and G, the number of Fs is the same as the number of Gs iff F and G can be put into one-one correspondence.

plicit definition".[23] In short, their claim involves deviating from Frege's own explicitly argued views on the nature of definitions: it takes a position on definitions similar to the one adopted by Hilbert when he classified various axioms in his *Foundations of Geometry* as definitions. Given the energetic and sustained attack that Frege leveled at Hilbert's position on definitions, one wonders if it is accurate to call the neo-Fregean's position "Fregean".[24] The neo-Fregean position is also non-Fregean in so far as it classifies Hume's Sentence as a definition (and not a logical law), thus contradicting Frege's explicit claim that the propositions of arithmetic are "laws of logic" ((Frege, 1959), p. 99). Furthermore, one wonders why one couldn't take the Peano Axioms themselves as forming an implicit definition (in the way Hilbert regarded some of his geometric definition)—thereby avoiding the necessity of proving Frege's Theorem.[25] Of course, using such an "implicit definition" (i.e. Peano's Axioms) in one's defense of Frege would call into question the view that the neo-Fregeans are defending a form of Logicism. (Recall Russell's comment about "theft over honest toil").

In any case, what I wish to take up, here, is the kind of issue I discussed above regarding the neo-intuitionists' philosophy of mathematics: Ask of the neo-Fregean account of arithmetic if it is an account of actual arithmetic or of some philosopher's model of arithmetic? If Frege's Theorem is to be used to justify the position that all the assertions of *actual arithmetic* are analytic, then one needs to carry out for actual arithmetic the kind of justification given for claiming that the theorems of second-order arithmetic are all analytic. Thus, one would need to show, among other things, how to translate the sentences and proofs of the neo-Fregean's second-order system of arithmetic into a natural language (since actual arithmetic is carried out and expressed in a natural language). But to carry out such a translation, the neo-Fregean's logical language needs to be given an appropriate interpretation: presumably the sort of interpretation I have called

---

[23] See (Hale & Wright, 2001), pp. 11–14.

[24] See ..(Chihara, 2004), Chapter 2, Section 1, for a detailed discussion of the dispute between Frege and Hilbert concerning the nature of definitions.

[25] For an excellent discussion of the neo-Fregean's views on implicit definitions and Hume's Principle, see John MacFarlane's "Double Vision: Two Questions about the Neo-Fregean Programme" (to be published in a special issue of *Synthese*, edited by Oystein Linnebo).

an "NL interpretation".²⁶ How would that be done?

If the theorems of the neo-Fregean's arithmetic are to have the features that Frege attributed to them, then the object variables of the system must have "the widest domain of all", since the theorems are taken by Frege to apply to everything thinkable. Thus, the "object" variables of the system range over not just some arbitrarily selected domain such as the totality of physical objects, but rather all "objects". But is there any such well-defined totality? Why might one have doubts that there is?

We know that Frege regarded the equator, the moon, sets, the direction of the axis of the earth, and extensions of concepts (of any order at all) to be objects. Can we be confident that the notion of "object" that Frege has described is sufficiently clear and definite to determine a well-defined totality of all "objects", especially given the Fregean doctrine that every extension (of a concept—no matter what its order) is an object. If axiom V of Frege's system is abandoned, what concepts have extensions? If we leave that question unanswered, is there not a certain fuzziness about the very notion of extension²⁷ and does not this fuzziness suggest a certain amount of indeterminacy in the domain of objects? Then do the "general logical laws" of the system, as interpreted in the way being suggested, have determinate truth values? If not, this would certainly undermine the claim that the arithmetical laws are all analytic truths—or even truths.

What about the second-order variables? What are they supposed to range over? For Frege, they were supposed to range over second –order concepts. But is there a well-defined totality of all second-order concepts? Do we have any clear and distinct ideas about what concepts exist and how to determine what concepts exist? Here, one cannot appeal to set theory in specifying the domains of the variables without incurring the suspicion that the justification of the analyticity of arithmetic is ultimately circular, presupposing not just logical laws but various truths and concepts of higher order mathematics. Nor can one appeal to theorems of model theory in giving the interpretation of the concept variables, without raising doubts about the claim that only general laws of logic and definitions are used in the proof.

---

²⁶ See (Chihara, 2004), pp. 34–6, for a discussion of NL interpretations.

²⁷ It should be noted that, after he learned of the inconsistency in his system from Russell, Frege expressed his doubts about the notion of 'extension of a concept', writing: "Every day I understand less and less what is really meant by 'extension of a concept' ((Frege, 1980), p. 139).

Suppose that one were able to show somehow that the natural language translations of the assertions of second-order arithmetic are all analytic. What would that show about actual arithmetic? Even if the neo-Fregean's translations of the theorems of their system of formal arithmetic are shown to be analytic, what would that show about the English sentences that express theorems of actual arithmetic? Certainly, it would be extremely implausible to hold that the neo—Fregean's version of '$2+2=4$' captures or gives the meaning of the English statement 'Two plus two equals four'. What then is the relationship between the neo-Fregean's version and the corresponding English statement? When ordinary working scientists or engineers assert arithmetical statements, are they asserting what the neo-Fregean's versions of the statements assert? Are they making assertions about "objects", "concepts", and "extensions" (in Frege's senses of those words)? Surely not. It thus needs to be spelled out in detail how Frege's Theorem allows us to infer that the truths of actual arithmetic are analytic. Limitations of space preclude my delving deeper here into these problems or in my exploring other similar neglected problems of 20th Century philosophy of mathematics.

What are the most important open problems in the philosophy of mathematics and what are the prospects for progress?

In the space I have left, I would like to discuss briefly an important problem in philosophy of mathematics that is still unsolved after over two thousand years of attempts at a solution. This problem, which I call "the diagnostic problem" of the "vicious-circle paradoxes", served as among the main motivations for Bertrand Russell's ramified theory of types, as well as his major work in philosophy of mathematics (written with Alfred North Whitehead) *Principia Mathematica*.[28]

What is this problem? A paradox, as I see it, is an argument that has premises that appear to be obviously true, that proceeds according to rules of inference that appear to be obviously valid, but that has a conclusion that is obviously false (say, a self-contradictory conclusion). Evidently, *something* in the argument appears to be the case even though it is in fact not the case, and the diagnostic problem of a paradox is to determine what in

---

[28] See (Chihara, 1973), Chapter 1, for details.

the paradox this *something* is. In other words, *the problem is to provide a correct diagnosis of the paradox*. What the philosopher would like to know is not only where we went wrong, but also *how* we were deceived into thinking that what is not the case is the case?

The diagnostic problem of a paradox is to be contrasted with what I call "the remediation problem" of the paradox: the problem of revising or repairing our rules of inference, our assumptions, or our system of concepts so as to block the type of reasoning that gave rise to the paradox, while preserving all (or at least most) of what was useful or important in the original set of rules, beliefs, or concepts).

Limitations of space prevent me either from going into the various semantic vicious-circle paradoxes that Russell attempted to resolve (and diagnose) with his theory of types or even from explaining why Russell thought that solving these paradoxes was so crucial for his project of setting mathematics on a firm logical foundation.[29] However, I can say that I agree with him that solving these paradoxes would significantly aid us in better understanding the nature of the set theoretical paradoxes that produced what many consider a serious crisis in the foundations of mathematics in the early Twentieth Century.[30] Due to limitations of space, I shall have to confine my discussion here to only one of these vicious-circle paradoxes: the Liar Paradox.[31]

Why do I think that the diagnostic problem of the Liar Paradox is important? From my perspective, any long-standing unresolved paradoxes is important. (Recall my earlier discussion of the nature of philosophy). In this case, the lack of an attractive diagnosis is a strong indication of a deep failure of philosophy to produce a coherent account of a key fundamental concept that has played, and continues to play, an important role in the philosophy of mathematics (as well as practically all aspects of science and everyday life)—viz . the concept of truth. A solution to this problem can

---

[29] See (Chihara, 1973), Chapter 1, for an explanation of the vicious-circle paradoxes that Russell considered, as well as of his reasons for placing such importance on solving the paradoxes.

[30] The appearance of the "antinomies in the basis of set theory" are described by A. Fraenkel and Y. Bar-Hillel as bringing about "the third foundational crisis" of mathematics ((Fraenkel & Bar-Hillel, 1958), p. 15). In (Kline, 1982), Chapter 9, Morris Kline entitles the chapter on the set theoretical paradoxes: "Paradise Barred: A New Crises of Reason".

[31] See (Chihara, 1979), Section 7, for a statement of this paradox.

be expected to have important implications for this important concept. A solution should also tell us something of importance about the set theoretical and mathematical paradoxes. The solution can even be expected to have far-reaching implications for other fundamental (and related) concepts of philosophy that figure in different versions of the paradoxes, such as reference and necessity.

Over the centuries, a great many different solutions to that paradox have been advanced, but it is safe to say that, as yet, no solution has won sufficient acceptance from the logical-philosophical community to warrant calling the paradox "solved". For the most part, the solutions proposed have taken the form of pinpointing some inference in the argument as invalid. In recent years, with the development of sophisticated advances in mathematical logic, there have been a number of mathematically complex analyses of truth that suggest possible "diagnoses" of the paradox–analyses that involve postulating logically intricate features of the truth predicate in virtue of which some inference in the argument can be blocked or seen to be invalid.[32] If such analyses were taken to provide us with solutions to the diagnostic problem, then I would judge them to be implausible, primarily because even young children can follow the reasoning of the paradox and can then be put in the grip of puzzlement at the demonstration of how one can apparently validly infer a contradiction from seemingly true premises. These children clearly have a sufficiently developed conception of truth to follow the reasoning of the paradox. That they have a conception of truth with the sophisticated mathematical features that most of these analyses postulate, however, strikes me as not credible. (The children I have in mind have had no higher mathematics-even calculus or elementary set theory–and have at best a naïve conception of infinite totalities). Thus, even if one were convinced that mathematically sophisticated thinkers have a complex concept of truth that has the features that some of these recent analyses presuppose, one would still need to understand what is going on in the minds of the mathematically unsophisticated thinkers, such as the children to whom I have posed the paradox. They clearly have a workable notion of truth sufficient to understand most sentences containing the word 'true' one is apt to hear in common everyday interactions with other people, and they clearly have a sufficient understanding of the reasoning of the

---

[32] See (Beall, 2007) for a nice overview of many of these analyses of truth.

paradox to be perplexed by it. So one would still need a diagnostic solution to the paradox under the form that these children would face. If such a solution could be found, then it would be natural to wonder if the solution applies equally well to the form of the paradox facing the sophisticated mathematician.[33]

What are the prospects for progress?

With the space I have left, I cannot hope to provide here anything like a convincing solution to the problem, but I would like to point the way to a type of solution I have found promising and indicate why I find such a diagnosis of the paradox to be attractive. My suggested diagnosis is that the common ordinary everyday concept of truth (one that is possessed by even the children referred to above) is *inconsistent* (or inconsistent with *known facts of reference*). In other words, I hypothesize that the crucial premises of the paradox do accurately reflect what is understood, or at least implied, by what one might call "the naïve concept of truth". Since according to the diagnosis, an accurate definition of the truth concept would be inconsistent, those possessing the concept will find the premises of the Liar to be incontestable, and there would be no need to find some fallacy underlying the reasoning of the paradox: the reasoning of the paradox would be faultless. This type of diagnosis attributes the "cause" of the paradox to the very concept of truth. The inconsistency is to be found in the concept of truth itself, and that is what makes the premises seem so convincing and undeniable: a premise like 'A sentence S is true iff what S says to be the case is the case' might plausibly be said to express in a vague intuitive way "what we mean by true".

Why do I find such a diagnosis attractive? Start with the fact that for over two thousand years, some of the best minds in philosophy and logic have tried, but failed, to find some clear fallacy in the reasoning.[34] Given that the reasoning of the paradox is simple and straightforward, how likely is it that there is some definite flaw in the reasoning that has been overlooked by so many bright

---

[33] To some extent, this sort of reasoning can be applied (with suitable adjustments) to the "contextualist approaches" to the paradoxes. See, in particular (Beall, 2007), p. 387, especially fn. 156. For a detailed discussion of contextualist approaches, see (see (Beall, 2007), Section 7.2).

[34] According to Alfred Tarski, the Liar Paradox 'tormented many ancient logicians and caused the premature death of at least one of them, Philetas of Cos' ((Tarski, 1969), p.66).

people? Is it not more plausible that there is no such fallacy to be uncovered?

Many philosophers find it very difficult to allow that such a fundamental and important concept as truth *can be* inconsistent. It has been even suggested to me that if our concept of truth were inconsistent, the importance of truth–the very significance that we impute to truth–would be an illusion. Why might one think so? The reasoning would go along the following lines: If the concept of truth were inconsistent, as it is being suggested, then the predicate 'is true' would be explicable by a formula or definition that is inconsistent (or inconsistent with known facts). Now it is held that:

> *an explicating formula or definition would determine whether a predicate (such as 'is true') applies to a sentence or statement x only if there is no valid argument from true statements and the explicating formula to the conclusion both that the predicate does apply to x and that the predicate does not apply to x.*

Thus, no inconsistent formula or definition could specify a genuine condition or extension, and hence such an inconsistent predicate could not be used to make any sort of genuine statement. More specifically, my inconsistency hypothesis would imply that no statement of the form 'X is true' could be meaningful — something that is clearly absurd.[35]

In response to this objection, I would like to consider an organization, Sec Lib, that determines its eligibility requirements for admission to the organization in the following way: To be eligible for membership, the candidate must be the secretary of one and only one organization C. Such a candidate must then satisfy the following formula:

> The candidate is eligible to join Sec Lib iff
> he/she is not eligible to join C

No problem is encountered with this rule until a young lady, Ms. Fineline, becomes the secretary of Sec Lib (and no other organization) and applies for membership in the organization. According

---

[35] I owe this argument to George Myro. See (Chihara, 1984) for more details. This argument is also closely related to Charles Parson's suggestion that the inconsistency view saddles ordinary speakers with "an incoherent conceptual apparatus" (see (Chihara, 1979), pp. 608-9, for my response to this suggestion).

to the rule, Ms Fineline is eligible to join Sec Lib iff she is not eligible to join Sec Lib.[36] From this inconsistent conclusion, one could infer anything at all. So can we then infer that the rules of eligibility cannot be inconsistent, since such a conclusion would imply that all statements of the form 'A is eligible to join Sec Lib' could not be genuine statements? Surely not. For the rules of eligibility clearly *are* inconsistent when applied to the one case in which Ms. Fineline applies for membership: the above italicized principle is simply incorrect. And to conclude that the rules of eligibility *could not* be inconsistent, on the grounds that such a conclusion conflicts with the fact that, prior to the problem arising with Ms. Fineline, we made perfectly acceptable statements about eligibility to join the organization—would itself be absurd!

Could one respond to my position by arguing that one can construct a valid argument from the inconsistency about Ms. Fineline derived above to the conclusion that *Ms. Jones*, who was earlier deemed eligible to join Sec Lib, *is not eligible to join Sec Lib* after all? No one with an ounce of common sense would draw such a conclusion, any more than someone would infer from such an inconsistency that the twin primes conjecture is true. No reasonable thinker would draw such an inference.

Thus, for all practical purposes, the rules do determine eligibility for all but one case, and except for that one case the truth of sentences of the form 'A is eligible to join Sec Lib' can be determined for any candidate who is secretary of one and only one organization C according to the formula:

The candidate is eligible to join Sec Lib iff he/she is not eligible to join C

Notice that, in the case of the Liar Paradox, the contradictory results appear only for very special sorts of sentences; for the vast majority of everyday assertions involving sentences in which 'is true' appears in predicate position, no problems occur. Like Sec Lib's eligibility rule, one would not run into trouble with the ordinary concept of truth except in somewhat extraordinary situations. For *practical purposes*, the naïve concept of truth works

---

[36]This is a version of the Sec Lib Paradox that I presented in (Chihara, 1979) and attributed to Frank Cioffi (who told me that he got it from some science fiction story). However, more recently, I have been informed by Roy Sorensen that it was published in the *American Mathematics Monthly* (Volume 47, Issue 7 (Aug-Sep, 1940), p. 474) by L. S. Johnson in note entitled "Another Form of the Russell Paradox".

fine.[37] The mere fact that one can construct a valid argument, from the inconsistency derived in the Liar Paradox, to the conclusion, say, that every US citizen is entitled to receive a million dollars from the US Treasury—is no good reason to expect to receive a million dollars from the US Treasury. Common sense should prevail in allowing reasonable applications of the truth predicate.

Regarding the way common sense responds to paradox, consider the case of the calculus when it was just being developed by Leibniz and Newton. The problems at the foundations of their views about the calculus are well-documented. That basic principles and concepts of the calculus were muddled and, in parts, even unintelligible or inconsistent was forcefully argued by such critics as Bishop Berkeley.[38] But such foundational facts did not prevent mathematicians from producing many important theorems and applications. Nor did problems at the foundations of the new theory prompt mathematicians to abandon the new theory or regard its theorems as meaningless. No one, so far as I know, questioned, say, the intelligibility and value of the Fundamental Theorem of the Integral Calculus. Most mathematicians were convinced that the foundational problems did not affect much of the central work being done and that the problems would be eventually resolved and corrected. Thus, it was left for later mathematicians to redo the foundations of analysis so as to eliminate the many muddles and inconsistencies, while preserving its unquestioned valuable core.

Turn now to the case of naïve set theory. At one time, the following two axioms were thought by many mathematicians to give the essential nature of sets:

Abstraction Axiom: For every property or condition, there is a set whose members are just those objects that have the property or satisfy the condition.

Extensionality Axiom: For any set A and any set B, if A and B have the same members, then $A = B$.

Before the appearance of the paradoxes, when asked why one should accept the axioms, the reply could be given that these axioms express the essential nature of sets: "That is what sets

---

[37] The reader can find additional details of my response in (Chihara, 1984).

[38] See (Kline, 1982), Chapter 6, for an indication of the mess at the foundations of the calculus in the early years of the development of analysis.

are"; "That is what we mean by 'sets'"; "These axioms implicitly define the term 'set'". As support for such views about the axioms, note that Hilbert claimed that certain of the axioms of his geometry "define the concept of 'between'" and certain others "define the concept of congruence" (see (Chihara, 2004), p. 29). And it has become common for philosophers and mathematicians to take the axioms of a formal system to define a type of structure. Of course, from the above axioms of naïve set theory, one can derive Russell's Paradox. So it is easy to see why one might now hold that the "naïve conception of set", possessed by those who accepted the above axioms, was faulty: the very conception of set was inconsistent.[39] But should we suppose that such a conceptual inconsistency would completely undermine the value of all the work that had been done in set theory? One found ways of retaining the mathematically essential features of the concept, while eliminating some features that gave rise to the paradox. In other words, mathematicians found many alternative ways of giving relatively satisfactory solutions to *the remediation problem* of the set theoretical paradoxes.

Similarly, even if one were to accept the kind of inconsistency diagnosis of the Liar being suggested, I see no reason for concluding that all judgments of truth we have made would be undermined. Such a pessimistic view is no more warranted than the view that the confusions about the fundamentals of the calculus rendered worthless all the work of the early pioneers of analysis.

From the above perspective, one can see one important advantage of the present inconsistency proposal that most of the "solutions" now being proposed lack: once one accepts the inconsistency diagnosis, the various proposals for analyzing truth or our rules of reference can be regarded as alternative ways of responding to the remediation problem. Thus, there would be no need to select from this collection of analyses "the correct one". The proposals could be regarded as alternative ways of developing new concepts and rules that avoid the contradictions, while preserving what is most important of our present system. There need not be a single best way, but many alternate ways, each having its own advantages and strengths. This would be similar to the situation we had in set theory, after the discoveries of the set theoretical paradoxes:

---

[39] Cf. "The antinomies show that the naïve concept of set as appearing in Cantor's "definition" of set ... cannot form a satisfactory basis of set theory ..." ((Fraenkel & Bar-Hillel, 1958), p. 19).

there was no need to single out as "the correct one" any one of the many set theories that had been formulated since the paradoxes were discovered. Researchers developing paradox-free concepts of truth should find it liberating not to have to justify their analyses as providing us with solutions to the diagnostic problem.

References

Bar-Hillel, Y., Posnanski, E. I. J., Rabin, M. O., & Robinson, A. (Eds.). (1966). *Essays on the Foundations of Mathematics.* Jerusalem: at The Magnes Press, The Hebrew University.

Beall, J. C. (2007). Truth and Paradox: A Philosophical Sketch. In D. Jacquette (Ed.), *Handbook of the Philosophy of Science: Philosophy of Logic* (pp. 325-410). Amsterdam: North-Holland Elsevier.

Beth, E. W. (1959). *The Foundations of Mathematics.* Amsterdam: North-Holland Publishing Company.

Burgess, J., & Rosen, G. (1997). *A Subject With No Object: Strategies for Nominalistic Interpretations of Mathematics.* Oxford: Oxford University Press.

Chihara, C. S. (1973). *Ontology and the Vicious-Circle Principle.* Ithaca: Cornell University Press.

Chihara, C. S. (1979). The Semantic Paradoxes: A Diagnostic Investigation. *The Philosophical Review, 88,* 590–618.

Chihara, C. S. (1984). The Semantic Paradoxes: Some Second Thoughts. *Philosophical Studies, 45,* 223–229.

Chihara, C. S. (2004). *A Structural Account of Mathematics.* Oxford: Oxford University Press.

Dauben, J. W. (1988). Abraham Robinson and Nonstandard Analysis: History, Philosophy, and Foundations of Mathematics. In W. Aspray & P. Kitcher (Eds.), *History and Philosophy of Modern Mathematics* (pp. 177-200). Minneapolis: University of Minnesota Press.

Davis, M. (2005). *Applied Nonstandard Analysis* (Dover paperback ed.). Mineola, NY: Dover Publications.

Enderton, H. (1972). *A Mathematical Introduction to Logic.* New York: Academic Press.

Feferman, S. (1998a). Weyl Vindicated: Das Kontinuum Seventy Years Later. In *In the Light of Logic* (pp. 249–283). New York: Oxford University Press.

Feferman, S. (1998b). Why a Little Bit Goes a Long Way: Logical Foundations of Scientifically Applicable Mathematics. In *In the Light of Logic* (pp. 284-298). New York: Oxford University Press.

Field, H. (1980). *Science Without Numbers*. Princeton: Princeton University Press.

Fraenkel, A., & Bar-Hillel, Y. (1958). *Foundations of Set Theory*. Amsterdam: North-Holland Publishing Co.

Frege, G. (1959). *The Foundations of Arithmetic* (J. L. Austin, Trans. Second ed.). Oxford: Basil Blackwell.

Frege, G. (1979). Sources of Knowledge of Mathematics and the mathematical natural Sciences. In H. Hermes, F. Kambartel & F. Kaulbach (Eds.), *Gottlob Frege: Posthumous Writings* (pp. 267-274). Chicago: The University of Chicago Press.

Frege, G. (1980). *Philosophical and Mathematical Correspondence* (H. Kaal, Trans.). Chicago: University of Chicago Press.

Hale, B., & Wright, C. (2001). *The Reason's Proper Study: Essays toward a Neo-Fregean Philosophy of Mathematics*. Oxford: Oxford University Press.

Hale, B., & Wright, C. (2005). Logicism in the Twenty-First Century. In S. Shapiro (Ed.), *The Oxford Handbook of Philosophy of Mathematics and Logic* (pp. 166-202). New York: Oxford University Press.

Hatcher, W. (1968). *Foundations of Mathematics*. Philadelphia: W. B. Saunders Company.

Hellman, G. (1989). *Mathematics Without Numbers*. Oxford: Oxford University Press.

Hellman, G. (1998). Beyond Definitionism–But Not Too Far Beyond. In M. Schirn (Ed.), *The Philosophy of Mathematics Today* (pp. 215–225). Oxford: Oxford University Press.

Kline, M. (1982). *Mathematics: The Loss of Certainty* (Paperback edition ed.). New York: Oxford University Press.

Kneebone, G. T. (1963). *Mathematical Logic and the Foundations of Mathematics*. London: D. Van Nostrand.

Mancosu, P. (Ed.). (1998). *From Brouwer to Hilbert: The Debate on the Foundations of Mathematics in the 1920s*. New York: Oxford University Press.

Ostebee, A., Cambardella, P., & Dresden, M. (1976). "Nonstandard" Approach to the Thermodynamic Limit. *Physical Review A, 13*, 878–881.

Posy, C. (2005). Intuitionism and Philosophy. In S. Shapiro (Ed.), *The Oxford Handbook of Philosophy of Mathematics and Logic* (pp. 318–355). Oxford: Oxford University Press.

Robinson, A. (1996). *Non-standard Analysis* (Revised Edition ed.). Princeton: Princeton University Press.

Shoenfield, J. R. (1967). *Mathematical Logic*. Reading, Massachusetts: Addison-Wesley.

Sullivan, K. (1976). The Teaching of Elementary Calculus Using the Nonstandard Analysis Approach. *American Mathematical Monthly, 83*, 370–375.

Tarski, A. (1969). Truth and Proof. *Scientific American, 220*, 63–77.

Wang, H. (1962). *A Survey of Mathematical Logic*. Amsterdam: North-Holland.

Wilder, R. (1952). *Introduction to the Foundations of Mathematics*. New York: John Wiley & Sons.

Wright, C. (2001). On the Philosophical Significance of Frege's Theorem. In B. Hale & C. Wright (Eds.), *The Reason's Proper Study: Essays toward a Neo-Fregean Philosophy of Mathematics* (pp. 272–306). Oxford: Oxford University Press.

7
Mark Colyvan

Professor of Philosophy
University of Sydney, Australia

Why were you initially drawn to the foundations of mathematics and/or the philosophy of mathematics?

Like many philosophers of mathematics, I suspect, I started out in mathematics and drifted into philosophy. My undergraduate work was mostly in pure mathematics. I loved (and still love) algebra, functional analysis, topology, and complex analysis. While immersed in the many elegant proofs in these areas, I began wondering about the nature of proof and mathematical truth. I thus became interested in, and began studying, logic and philosophy as complements to my mathematics major. I completed a Bachelor of Science with honours in pure mathematics, with a thesis on the Dirichlet problem in potential theory. By the time I finished honours, I was fascinated by the many philosophical issues arising in mathematics. Some of those that attracted my attention were: (1) the existence or non-existence of mathematical objects, (2) the applications of mathematics in empirical science and (3) the question of proof and the status of axioms such as the axiom of choice.

Around this time I read Hartry Field's *Science without Numbers* (1980) and Penelope Maddy's paper 'Indispensability and Practice' (1992). After reading these two pieces, there was no turning back. I switched from mathematics to philosophy. I started a Ph.D. in the philosophy of mathematics at the Australian National University, where I was very fortunate to have Jack Smart as my supervisor. Jack's genuine, on-going fascination with science and mathematics, and his passion for learning has been an inspiration to me. Indeed, not long after starting my graduate work, Jack expressed interest in supervising me because he had always wanted to work on the philosophy of mathematics and then in his retirement, he finally had the time to come up to speed in this area. We thus worked though the contemporary literature – Maddy (1990), Field (1980, 1989), etc. – together. It was a wonderful experience and I am very pleased to say that Jack too has written up some of his thoughts on the philosophy of mathematics (Smart manuscript).

During my time as a graduate student I was fortunate enough to meet, and spend time with, Hartry Field, Penelope Maddy, and Mike Resnik. These three, in particular, have been enormously influential on my intellectual development. They set very high standards in their own work, providing shining examples of how good philosophy should be done and how such philosophy can genuinely advance debates in the philosophy of mathematics (and in mathematics itself). They were also very generous with their time and ideas. Mike Resnik, for instance, effectively steered me towards my thesis topic: a defence of the Quine-Putnam indispensability argument. (This is an argument that we ought to take mathematical objects to exist because of the indispensable applications of mathematics in empirical science (Quine 1981, Putnam 1979, Colyvan 2001).) Initially I thought Mike was crazy. A whole thesis on *one* argument! Surely, I thought, I'd need to do more than that. But Mike's advice was right on the money. Not only did he direct me to a *manageable* thesis topic, he directed me to a topic of considerable contemporary signigicance. Although my research interests have since broadened, I continue to work on the indispensability argument and other topics in the philosophy of mathematics. This is in part a result of the ongoing intellectual engagement with the likes of Hartry Field, Penelope Maddy, Mike Resnik, Jack Smart, and many others (indeed, many of the philosophers interviewed for this volume).

What examples from your work (or the work of others) illustrate the use of mathematics for philosophy?

Where do I begin? Mathematical methods are useful almost everywhere in philosophy. There are many obvious examples in the more technical areas of philosophy, such as philosophy of science, philosophy of mathematics, philosophical logic, decision theory, and formal epistemology. But there are also many less-technical areas where mathematics has been usefully employed to shed light on philosophical problems. Let me give an example of the latter from a paper I wrote with Jay Garfield and Graham Priest (Colyvan et al. 2005) on fine tuning arguments for the existence of an intelligent designer. (Also see some of Elliott Sober's terrific work on more general design arguments, for example, Sober 2002. On a related philosophy of religion theme, see Alan Hájek's (2003) fascinating discussion of Pascal's Wager—especially the discussion of the non-standard analysis angle.)

Design arguments draw attention to some feature of the world and suggest that the feature in question would be unlikely without the intervention of an intelligent designer. Various features of the world have been put forward as the focus of design arguments: the mechanical clock-like universe and complex biological structures. Biology was an especially popular focus for design arguments before Darwin's time (although sadly almost 150 years after the publication of *The Origin of Species* such arguments are still with us). Recently a physics-based version of the design argument has reared its head. This latter argument is usually known as *the fine-tuning argument*. Here the special feature of the universe that allegedly is in need of explanation is the existence of carbon-based life. It is noted that the universes seems to be fine tuned for the emergence of such life: had things been just slightly different, carbon-based life could not have evolved. For example, (for reasons that need not concern us here) had the fine-structure constant been even a few percent from its actual value, there would be no carbon molecules and hence there would be no carbon-based life. So, the argument continues, the universe as we find it (i.e. *with* carbon-based life) is improbable and this improbable state of affairs requires an explanation. The final move in the argument is to invoke an intelligent designer to provide the required explanation. This designer is presumably predisposed to value carbon-based life and is usually taken to be something like the Judeo-Christian god. Putting aside the obvious dubiousness of the last move, consider the argument up to the conclusion that an intelligent designer exists. Almost every premise and inference of even this part of the argument can be challenged, but one rather interesting mathematical challenge is right at the start, at the move from the universe being fine tuned to the universe being improbable.

Let's look at this move a little more closely. Presumably the idea is that the fine-structure constant, say, could have taken a value from a large range of real numbers and yet only a small subinterval in this range will permit carbon-based life. But before we can move to assigning even qualitative probabilities, we need to know something about the probability distribution in question. Of course, if the proponents of this argument are supposing that the fine-structure constant could have taken any real number as its value and all such values are equally likely, then the argument is fatally flawed. There simply can be no uniform distribution over an unbounded set like the reals (or even the positive reals). So how do we get to the improbability claim? The claim must be that the distribution in question is not uniform or that it has compact support. But that's not enough. We also require that the shape of the distribution is such that it delivers a low probability to the carbon-based-life-permitting interval. Typically, none of this is argued for by proponents of the fine-tuning argument. Indeed, some of the discussions in this area are mathematically very naïve, bogging down in extended and misguided treatments of how to give frequency interpretations of the probabilities in question, rather than adopting the standard measure-theoretic approach.

These mathematical reflections on the problem don't provide a knock-down argument against all fine tuning arguments, but they do block many of the less careful presentations. These reflections also make it clear what proponents of these arguments need to demonstrate. I can't help stressing how difficult their task is though. They need to argue that the distribution in question has exactly the right shape to ensure that the carbon-based-life-permitting interval has low (but non-zero) probability, and that this probability is higher under the assumption of an intelligent designer. Moreover, they need to show that the intelligent designer is the best explanation of the presence of carbon-based life in the universe. The latter, of course, is highly non-trivial. To take one often-overlooked hypothesis that strikes me as very plausible: the distribution in question is such that it has most of its density over the carbon-based-life-permitting interval. Given that we know next to nothing about this distribution (hence the aforementioned assumption of the distribution being uniform in many naïve presentations of the fine-tuning argument), we can use the evidence that there is carbon-based life to support my suggested alternative hypothesis about the shape of the distribution. A little bit of mathematics and these design arguments start to crumble.

What is the proper role of philosophy of mathematics in relation to logic, foundations of mathematics, the traditional core areas of mathematics, and science?

My view about the relationship between science and philosophy is a naturalistic one, where I take the role of philosophy as that of helping to understand science from within the scientific enterprise. This does not, for example, mean that philosophy is merely a powerless public servant, rubber stamping all and only the pronouncements of our current best science. Philosophy has an important role here, providing details of a plausible epistemology (for example, by providing an account of scientific confirmation), shedding light on metaphysical issues (for example, about the nature of causation), and critiquing science and subjecting it to scrutiny (for example, by examining the philosophical underpinnings of the statistical methods used in various branches of science). There is nothing new in any of this. This view of philosophy's relationship to science goes back a long way (at least to Russell) and was made famous and elegantly articulated by W.V.O. Quine in a number of places (for example, in Quine 1981).

How this naturalistic attitude plays out in relation to mathematics and logic is a little more complicated, but the basic idea is the same. The philosophical enterprise in relation to logic and mathematics is to provide a satisfying epistemology and, amongst other things, to make sense of current debates about the appropriate logic and new candidate axioms for set theory. Penelope Maddy (1997), in particular, has done some important work on the 20th Century debates over the axioms of ZFC set theory and the contemporary debate over new axioms. She takes a naturalistic approach and, in passing, shows how such an approach can genuinely advance both philosophy and mathematics.

What do you consider the most neglected topics and/or contributions in late 20th century philosophy of mathematics?

Let me start with a topic that was neglected but has since become one of the main foci of contemporary philosophy of mathematics: the applications of mathematics. The original motivation for much of the work on the applications of mathematics was the Quine-Putnam Indispensability Argument. Hartry Field, Penelope Maddy, and other critics of this argument were led to examine, in detail, specific applications of mathematics in science. This work has helped focus the debate about the indispensability argument, but it has also been interesting in its own right. Irrespective of your metaphysical leanings, an important fact about mathematics is that it finds widespread and diverse applications in science: from group theory in fundamental particle physics, to differential equations in population ecology. The recent work on applications, begun by Field (1980, 1989) and Maddy (1997), and, in a different context, by Steiner (1998) and Wigner (1960), has resulted in what might even be considered a new area—the philosophy of applied mathematics. This area has been very fruitful and has helped shift, or at least broaden, the focus of the philosophy of mathematics from the more traditional, foundational questions about pure mathematics. These traditional questions are, of course, still interesting and deserving of attention, but focusing on pure mathematics and its foundations, in isolation, I think, is to miss an important part of the story. Pure mathematics finds applications elsewhere in mathematics and in science. And as I have argued elsewhere (Colyvan 2001a), philosophers of both realist and anti-realist stripes need an account of applied mathematics, so this is no place for passing the buck on the required philosophical work. Happily, this work on applications is now well and truly under way and forms a significant thread in contemporary philosophy of mathematics. I should add that this work on applications also has interesting connections with related work in the philosophy of science on idealizations in scientific models.

Now to one of the neglected topics in the philosophy of mathematics. The old debate about the correct logic for mathematics never moved past intuitionism versus classical logic, but in the context of inconsistent mathematical theories there is the issue of paraconsistent logic versus the rest. A paraconsistent logic is one in which *ex contradictione quodlibet* (or "explosion") fails. That is, unlike classical logic, a paraconsistent logic does not allow the derivation of an arbitrary sentence from an arbitrary contradiction. Such logics have important applications, such as in modeling inconsistent agents and as insurance in large, possibly inconsistent, data sets (so that the contradictions don't spread and trivialise the data set). But there are also important applications in the philosophy of mathematics. Let me briefly mention a couple.

There have been times when we've been forced to work with inconsistent mathematical theories (for example, naïve set theory and, arguably, the early calculus). What is more, we were forced to work with these theories at times when they were known to be inconsistent. The first point to note is that using classical logic and taking an explicit statement of the contradiction as an assumption, a proof of any mathematical sentence can be derived in a few lines. But, perhaps unsurprisingly, such cheap proofs were not taken to be legitimate. What does this suggest about the logic of mathematics? Is it paraconsistent? Or is the logic classical but with further pragmatic constraints imposed to avoid the cheap proofs just mentioned? It is also interesting to note that the two contradictory theories I just mentioned (the early calculus and naïve set theory) were major mathematical theories with wide-ranging applications. This gives rise to another question: how is it that an inconsistent theory can be successful in its applications?

Another application of paraconsistent logic in the philosophy of mathematics is to model unashamedly inconsistent theories, such as those discussed by Bob Meyer and Chris Mortensen (1984), Chris Mortensen (1995), and Graham Priest (1997, 2000). Such theories have a number of virtues: they are non-trivial (in the sense that despite their inconsistency, not every sentence is provable), and they do not need to be incomplete. (After all, Gödel's incompleteness theorems, show that we must choose between consistency and completeness.) Even if we prefer our mathematics to be consistent (and we can find reasons to suppose that the theories in question are in fact consistent), inconsistent mathematics theories still have considerable interest. For example, the mere fact that inconsistent mathematical theories can find wide-spread applications in empirical science suggests that any philosophical account of the applications of mathematics in science had better not lean too heavily on consistency to do the work. It would seem that consistency is neither necessary nor sufficient for applicability. Thus far, only a few people (mostly paraconsistent logicians) have worked on these issues and they are yet to find their way into mainstream philosophy of mathematics. But I'm hoping that will change!

What are the most important open problems in the philosophy of mathematics and what are the prospects for progress?

There are many important questions in the philosophy of mathematics and almost all of them are open: how does an (apparently) a priori discipline like mathematics find applications in empirical science?; what is the appropriate logic for mathematics?; what status should we give to non-trivial inconsistent mathematical theories?; what is the appropriate attitude to have towards the posits of mathematics?; is mathematics a science of structures? (see Resnik, 1997 and Shapiro, 1997); what should count as an appropriate standard of rigour and can picture proofs meet such standards? (see Brown 1999); and many others. Here I'll say a little about the issue of explanation in mathematics and its relationship to theories of explanation elsewhere in science. This issue has received relatively little attention from philosophers of mathematics (although see Baker 2005, Colyvan 2007, Mancosu 2001, Resnik and Kushner 1987, and Steiner 1978).

It is generally thought that some proofs of theorems are explanatory while others are not. Indeed, two different proofs of the theorem may differ in their explanatory power. The idea is that anyone who properly understands an explanatory proof of a theorem, understands *why* the theorem in question is true. While a full understanding of a non-explanatory proof merely convinces one that the theorem is true, but does not give any insight into why it is true. It is sometimes thought that this is just another way of marking the constructive–non-constructive distinction, but that's not right. Some non-constructive proofs are explanatory and some constructive proofs fail to be explanatory. The problem for the philosophy of mathematics is to give an account of the notion of explanation at work here. It clearly can't be any causal account of explanation, such as those dominating discussions of explanation in the philosophy of science literature. According to the causal account of explanation, providing an explanation of an event is (roughly) to trace the event's causal history. No matter what your theory of causation is, a mathematical truth cannot be explained in terms of causal histories. There is no causal chain that has as its end point Green's Theorem. Proofs, whatever they are, are not narratives about causal histories.

Does this mean that mathematical explanation is different in kind from explanation elsewhere in science? I don't think so. Apart from a prima facie case (driven by simplicity considerations) that there ought to be just one account of explanation, irrespective of the domain in which the explanations arise, there is a more substantial reason for insisting on a unified account of explanation. Often the explanation of a mathematical result, such as Green's Theorem, can "spill over" into empirical science. There are many physical systems (for example fluids flowing through a region of space) that can be modelled by differential equations and the explanation of certain features of this system will be provided by the relevant mathematical results. (For example, the proof of Green's theorem will provide the explanation of the relationship between the flow in the interior of the region and on the boundary of the region.) So if we are to treat mathematical explanation as different in kind from explanation in physics, we will also, it seems, need to countenance two kinds of explanation in physics. That, of course, is not to say that such a piecemeal account of explanation is not right or can't be made to work, but it does suggest that the possibility of a unified account is worth exploring.

One initially promising candidate for a satisfying account of explanation in mathematics is the unification account of explanation (Kitcher 1981, Friedman 1974). According to this account, an explanation is just a unification of the phenomenon in question. Think, for example, of how one of the jewels of complex analysis, the Residue Theorem, generalizes Cauchy's Integral Theorem and provides a means for calculating the integrals of many real-valued functions. Or to take another example, consider how, Euler's formula, $e^{ix} = \cos x + i\sin x$ (where $i = \sqrt{-1}$ and x is any real number), unifies analysis and trigonometry. Indeed, our understanding of both analysis and trigonometry is significantly enhanced by such results in complex analysis. According to the unificatory account of explanation, these unifications are genuinely explanatory. Although there are various objections to the unification account of explanation, it seems the one account with any chance of success in mathematics and hence the only account with any chance of yielding a unified account of explanation across both mathematical and other scientific domains. Or so it seems to me, at least. In any case, I think that there is a great deal of interesting work yet to be done on mathematical explanation and its relationship to explanation in the broader scientific context.

References

Baker, A. 2005. 'Are there Genuine Mathematical Explanations of Physical Phenomena?', *Mind*, 114, 223–238.

Brown, J.R. 1999. *Philosophy of Mathematics: An Introduction to the World of Proofs and Pictures*, Routledge, New York.

Colyvan, M. 2001. *The Indispensability of Mathematics*, Oxford University Press, New York.

Colyvan, M. 2001a. 'The Miracle of Applied Mathematics', *Synthese*, 127, 265–278.

Colyvan, M. 2007. 'Mathematical Recreation Versus Mathematical Knowledge' in M. Leng, A. Paseau, and M. Potter (eds.), *Mathematical Knowledge*, Oxford University Press, forthcoming.

Colyvan, M., Garfield, and J.L. Priest, G. 2005. 'Problems with the Argument from Fine Tuning', *Synthese*, 145, 325–338.

Field, H. 1980. *Science Without Numbers: A Defence of Nominalism*, Blackwell Publishers, Oxford.

Field H. 1989. *Realism, Mathematics and Modality*, Blackwell Publishers, Oxford.

Friedman, M. 1974. 'Explanation and Scientific Understanding', *Journal of Philosophy*, 71, 5–19.

Hájek, A. 2003. 'Waging War On Pascal's Wager', *Philosophical Review*, 113, 27–56.

Kitcher, P. 1981. 'Explanatory Unification', *Philosophy of Science*, 48, 507–531.

Maddy, P. 1990. *Realism in Mathematics*, Clarendon Press, Oxford.

Maddy, P. 1992. 'Indispensability and Practice', *Journal of Philosophy*, 89, 275–289.

Maddy, P. 1997. *Naturalism in Mathematics*, Clarendon Press, Oxford.

Mancosu, P. 2001. 'Mathematical Explanation: Problems and Prospects'., *Topoi*, 20, 97–117.

Meyer, R.K. and Mortensen, C. 1984. 'Inconsistent Models for Relevant Arithmetic', *Journal of Symbolic Logic*, 49: 917–929.

Mortensen, C. 1995. *Inconsistent Mathematics*, Kluwer, Dordrecht.

Priest, G. 1997. 'Inconsistent Models of Arithmetic Part I: Finite Models', *Journal of Philosophical Logic*, 26, 223–235.

Priest, G. 2000. 'Inconsistent Models of Arithmetic Part II: The General Case', *Journal of Symbolic Logic*, 65, 1519–1529.

Putnam, H. 1979. 'Philosophy of Logic' in *Mathematics Matter and Method: Philosophical Papers* Vol. I, second edition, Cambridge University Press, Cambridge, 323–357.

Quine, W.V. 1981. 'Five Milestones of Empiricism' in *Theories and Things*, Harvard University Press, Cambridge MA, 67–72.

Quine, W.V. 1981a. 'Success and Limits of Mathematization' in *Theories and Things*, Harvard University Press, Cambridge MA, 148–155.

Resnik, M.D. 1997. *Mathematics as a Science of Patterns*, Clarendon Press, Oxford.

Resnik, M.D. and Kushner, D. 1987. 'Explanation, Independence, and Realism in Mathematics', *British Journal for the Philosophy of Science*, 38, 141–158.

Shapiro, S. 1997. *Philosophy of Mathematics: Structure and Ontology*, Oxford University Press, New York.

Smart, J.J.C. Manuscript. 'Mathematical Realism', paper presented at the "Truth and Reality" conference at the University of Otago, Dunedin, New Zealand.

Sober, E. 2002. 'Intelligent Design and Probability Reasoning', *International Journal for the Philosophy of Religion*, 52, 65–80.

Steiner, M. 1978. 'Mathematical Explanation', *Philosophical Studies*, 34, 135–151.

Steiner, M. 1998. *The Applicability of Mathematics as a Philosophical Problem*, Harvard University Press, Cambridge MA.

Wigner, E.P. 1960. 'The Unreasonable Effectiveness of Mathematics in the Natural Sciences', *Communications on Pure and Applied Mathematics*, 13, 1–14.

8

E. Brian Davies

Professor of Pure Mathematics
University College London, UK

Why were you initially drawn to the foundations of mathematics and/or the philosophy of mathematics?

I first became interested in philosophy as a result of reading some of my father's books when I was in my early teens. He was a great admirer of Bertrand Russell and I read "A History of Western Philosophy" with great interest. Russell's extraordinary eloquence and his breadth of knowledge opened my eyes to ideas that I previously had no notion of. Although I was not to write anything in the field for several decades, I read books about science avidly from an early age, particularly those relating to evolution and consciousness.

During the 1950s an immensely popular series of television programmes featured Julian Huxley, A J Ayer, Jacob Bronowski and a number of other major public figures who discussed moral philosophy and related issues (but not party politics), at a level that is inconceivable for today's media. I lapped it up, accepting, as everyone did in those days, the deference that intellectuals such as these received.

During the 1970s I was loosely associated with a group of mathematical physicists trying to build a constructive quantum field theory. This subject was a development of axiomatic field theory. It had no overt connection with constructivism or intuitionism in the philosophical sense, but I observed that those working in the field were relying on those aspects of operator theory that had constructive versions, and that they were forced to retrace their steps whenever they used a proof based on compactness arguments. I also observed that the abstract theory of distributions, formulated in terms of topological vector spaces, was losing influence by comparison with elliptic regularity methods based on Sobolev space estimates, which were much more constructive.

When a degree in Mathematics and Philosophy was set up in Oxford I was one of the few College tutors willing to accept students in the field – most thought that such joint degrees could not be taken seriously. I went to King's College in 1981, and attended the Philosophy of Science seminar there, irregularly at first but gradually more frequently. The organizer of the seminar, Donald Gillies, who happened to live close by in Dulwich, befriended me and we occasionally met by chance on trains, and no doubt amazed our fellow passengers by our intense conversations on philosophical topics. I owe him a tremendous amount for his constant encouragement.

Karl Popper's books provided another influence on my philosophical development, in spite of the fact that I came to disagree with him on a number of issues. His ideas about refutation were very illuminating in many contexts, but I eventually decided that their scope was not as great as he claimed. Popper simply ignored a huge amount of biological knowledge, such as that in Hooke's Micrographia, that only depended on looking down a microscope to see things that nobody had any inkling of before. It would be anachronistic to accuse him of ignoring what has been learned about the planets by sending space probes to look at them, but many other discoveries owe nothing to his theory of knowledge. It is rather amusing that his reputation among scientists seems unassailable in spite of the many doubts that philosophers have expressed about his advocacy of refutation as the method of scientific investigation.

My active involvement in philosophy resulted from a steadily increasing conviction that (some) philosophers of mathematics were spending too great a proportion of their efforts on a very narrow range of topics, particularly logic and set theory. (Discussions about probability and quantum mechanics provided an occasional break.) I got to dread seeing yet another discussion of Gödel's theorems and their importance, when I knew that they had had almost no relevance to the work of most mathematicians. There were a few exceptions, such as Lakatos, but not enough. I felt that the philosophy of mathematics should be grounded in what the great majority of mathematicians did in their work, and most philosophers of mathematics gave little impression that they were interested in this.

What examples from your work (or the work of others) illustrate the use of mathematics for philosophy?

Mathematics provides philosophers a subject whose philosophical status has been contentious for many centuries. Wrongly so in my opinion, but nevertheless it has been. Most mathematicians are not very interested in philosophy, because they understand that it will not help them to prove their theorems. However, that should not be taken as a criticism of philosophy – paper manufacturers provide essential support to composers of music, but the reverse is only true in an extremely indirect way. Mathematicians enjoy reflecting on the nature of their subject every now and again, but it would be difficult to quantify the benefit of this.

My response above is perhaps too negative, because my philosophical predilections had a direct effect on my choice of research. But this is very unusual. I read Bishop's book on constructive analysis in 1967 when I was finishing my DPhil in Oxford and was particularly open to outside influences. It persuaded me that my dislike of the heavy reliance on the axiom of choice in two books of Dixmier on C*-algebras was reasonable; this eventually led me to abandon the study of operator algebras. As it happened George Mackey was visiting Oxford at that time and I heard of his challenge to find a non-commutative version of measure theory. I was ideally placed to solve this problem and to obtain my first major mathematical results by inventing a somewhat more constructive version of operator algebra theory called the theory of Sigma*-algebras. The fact that I (deliberately) made no mention of my constructivist leanings at that time (1968/70) was surely important in gaining acceptance of this work. Some others who were much influenced by Bishop's work were not so fortunate. However, I did not realize that it was to be one of the ingredients in my philosophical writings many years later.

In the 1990s I was reminded of these issues when I started to study problems in spectral theory and numerical analysis going under the name of pseudospectral theory. This subject investigates the gap between existence and computability (in practice, not in a world with indefinitely expandable resources) for eigenvalues of large matrices, and more generally for the spectra of linear operators. These questions had been neglected for decades because of the laborious character of numerical calculations and the lack of interest of most mathematicians in computation. Although I might have been attracted into the field partly because of ideas that I had picked up from Bishop, he himself emphasized that he was not interested in the feasibility of computations.

Pseudospectral theory has its source in numerical analysis, but it has been taken up by 'pure' mathematicians, who have realized that it suggests a number of previously totally unsuspected theorems about non-self-adjoint operators. The spectra and pseudospectra of various families of matrices, Toeplitz operators and differential operators have been determined, and there is also a growing pseudospectral analysis of various operators in the semi-classical limit. It may become a text-book example of a paradigm shift, in which the very special properties of self-adjoint operators stop being regarded as the standard against which other operators are measured.

What is the proper role of philosophy of mathematics in relation to logic, foundations of mathematics, the traditional core areas of mathematics, and science?

Before answering this question, I should state the obvious, that philosophy is a multi-faceted subject, whose practitioners have many goals. The disagreements between philosophers are often profound, with the result that comments about "philosophers" as a group cannot really make sense. I hope that my remarks below are understood as referring to particular trends in the subject, which will no doubt be recognized by experts.

The obsession with foundations (by those who are so obsessed) is a hang-over from the first part of the twentieth century. It only appears to be the central issue if one focuses on the final product – mathematics as it appears in journals and monographs – and ignores the process by which it comes into existence, which is just as interesting. It could be said that understanding the creative process is the subject matter of the sociology of mathematics, but such barriers between subjects are artificial. Insisting on a strict separation between the process of creating mathematics and mathematics itself is often associated with a value judgement, namely that mathematics should not be considered as a part of human culture.

Lakatos was of course well aware of this, but Karl Popper supported the idea that the creative impulse was not a proper subject for philosophical investigation. By maintaining that mathematics should be studied in isolation from the context in which it was created, one creates an insoluble paradox. If one believes that mathematics is not created by mathematicians, one is likely to conclude that it must be discovered, and hence must already exist in some ethereal Platonic world. But when one tries to analyze how mathematicians could be aware of that world no sensible answer is to be found. The idea of Penrose and Hameroff that it might be the result of microtubules in the brain magnifying the effects of some currently unknown physics is still too vague to be assessed, and has very little independent support.

The superficiality of the Platonic viewpoint is exposed by comparing it with current analyses of natural language, which is just as extraordinary as mathematics in its proper domain of applicability. Research into language involves historical and cultural comparisons, studies of infant development, brain scanning techniques and linguistic analysis, all of which can be developed in great detail. Nobody thinks it worthwhile to mention Plato's theory of ideal forms for physical objects (such as beds and tables) in this context.

Plato, Kurt Gödel, Roger Penrose and Alain Connes often refer to reason as if it was a sense organ. This is superficially attractive, but as well as being an obvious category error, it makes no sense in evolutionary terms. The ability to observe the Platonic world would have been no use to anyone until the last five thousand years, and it is highly implausible that such a bizarre sense organ could have evolved since then, particularly when its possession brings no particular reproductive advantage and most members of the species appear not to possess it. On the other hand cultural evolution, the development of skills and the co-option of abilities to achieve new goals provide a straightforward explanation if one considers mathematics to be a cultural construct. From this perspective Peano arithmetic is a generalization of our experience of counting, which itself is the result of a combination of the innate ability of all mammals to recognize very small numbers and our remarkable language skills, also genetically determined. This account seems not to be acceptable to some people, because it does not fit their Platonic model, in which it is supposed from the start that arithmetic is independent of the existence of the human species.

Although arithmetic is a human construct, we are correct in saying that dinosaurs (generally) had four legs each, before the human species existed. Once we introduce a classification scheme which regards fore limbs and hind limbs as similar types of entity, in spite of their different anatomy, and use a language which includes numbers, the statement about dinosaurs follows from the evidence that we have. Human beings cannot step outside this language even if they believe that a dinosaur was no more than a temporary aggregation of a somewhat indefinite number of atoms. The only people who have the option of denying the facts about dinosaurs are those who do not accept the reality of the remote past, e.g. creationists. By accepting its reality, we commit ourselves to describing it using our own language, suitably adapted.

Those who place mathematics in a cultural context are often accused of being relativists. Some of them may be, but the proof of a theorem can be just as definite from this viewpoint as it is for Platonists. Nobody suggests that other inventions (such as aircraft, spectacles and computer programs) only have the properties that they do by convention. An invented object may well have properties and applications unanticipated by its inventor, but this does not need a special explanation. Of course one might invent other things, such as Jordan algebras as well as C*-algebras, or whist as well as bridge, and then they then have different and equally objective properties. The more tightly specified the rules governing a concept, the more one should expect there to be agreement about the outcome of applying those rules. The task is to produce sets of rules that are very precise and have very rich consequences. This is not easy, but it is not wholly astonishing that there has been some success after more than two thousand years of effort in mathematics.

Platonists presumably believe that the unique factorization theorem for Gaussian integers (to pick an example at random) was already true before Gauss introduced the relevant definitions around 1830, but those who think that mathematics is a part of human culture would either deny this or not understand what the assertion is supposed to mean. If it means that the invention of the Gaussian integers and subsequent proof of the theorem could have happened earlier, this is either a historical claim or a platitude. If it is something more profound, there seems to be no type of evidence that could settle the matter one way or the other. Either one is a believer or one is not.

I have often been amused by the confident assertion that modern (pure) mathematics depends on ZFC, even though very few mathematicians have ever attended a course or read a book that mentions the foundations of set theory or logic. When I make this comment, one response is an assertion to the effect that mathematicians need not be aware of the basis of their work, any more than motor cars are aware that they use petrol or planets are aware of Newton's laws of motion. The problem with this is that it does not explain what generations of mathematicians could have been doing before 1900. To name one example, Cauchy's groundbreaking work on functions of a complex variable in the 1820s could not have depended on ZFC, and did not need any modifications as a result of the introduction of ZFC. In spite of that, it is one of the towering achievements of mathematics, used in a huge range of different fields.

Similarly the development of the calculus by Newton and Leibniz took place three hundred years before anyone had given a proper definition of the real number system. This does not imply that the latter was worthless – indeed it was a major landmark of late nineteenth century mathematics – but in retrospect its primary significance was that it allowed mathematicians to start the study of irregular functions, and eventually fractal geometry, not that it provided a firm foundation for earlier discoveries. Of course, at the time the latter was the motivation, but with hindsight one can take a different view.

The aspect of Popper's writing that appealed to me most was his insistence that his World 3, the world of human cultural constructs, must be regarded as having objective reality, because of its undeniable influence on physical events. (I am less happy with his discussion of World 2, the world of subjective experience.) I found his discussion of this very illuminating and used it as a basis for my antireductionist and pluralist views about mathematics and science.

I see in World 3 the antithesis of the dogmatic reductionism of those physicists who claim that ultimately one can base all explanations of physical phenomena on scientific laws expressed using mathematical equations. The exclusive focus on phenomena that could be described in mathematical terms using only efficient causes led to the spectacular growth of physics over the last four centuries. Nevertheless it excludes almost everything about our thoughts and social interactions that we find interesting. Eliminative materialism as described by the Churchlands solves this problem by denying the existence of beliefs, wishes, intentions etc., and I will not pretend to have much sympathy with it. The neural framework that underlies our human thought processes may one day be fully understood in physical terms, but the contents of our thoughts are so complex that we will probably only ever be able to describe them in terms of 'folk psychology'. There is no evidence that our use of concepts such as beliefs is purely a matter of social conditioning as the Churchlands imply. Indeed, the universality of such concepts over all cultures and times strongly suggests that they are innate.

A pluralist would say that one might be able to give several different explanations of the same event without conceding that one of them is necessarily more fundamental than the others. For example, if one wants to understand why the level of CFCs in the Earth's atmosphere increased steadily until about 1990, after which it stabilized and started gradually to decrease again, physics and chemistry provide little insight, even though everything that happened during this period was in conformity with physical laws. One must instead make reference to the Montreal Protocol of 1987, the political process associated with it and the intentions of those signing the Protocol, all of which are well documented as politico-historical events. This is not an exceptional case, but the norm for most of what happens around us. Pluralism allows two different accounts of our thought processes, one in terms of the physical structures in our brains and the other based on 'folk psychology', to co-exist if both are found useful, as they evidently are.

What do you consider the most neglected topics and/or contributions in late 20th century philosophy of mathematics?

Philosophers should take more notice of developments in computational complexity, which are making the endless past discussions of Turing machines look increasingly irrelevant. The possibility of solving a problem given a machine with indefinitely expandable memory and an indefinite time to work in may be of interest to logicians, but it has little to offer most other people. The travelling salesman problem is just one instance of a much more important class of question – about which problems can be solved in polynomial time. Indeed even this is not enough: the development of body scanners was dependent of the discovery of the Cooley-Tukey algorithm, which enabled Fourier transforms to be calculated in times of order $n \log(n)$ rather than n^2. Many number theorists are now being paid by Governments to try to find feasible ways of factorizing numbers with a few hundred digits – a problem that everyone agrees is trivial for a sufficiently (i.e. inhumanly) patient person with a Turing machine. If mathematicians find such issues important, then philosophers should take them on board too, or condemn themselves to discussing an ever smaller part of the subject.

A growing number of important mathematical theorems depend on rigorous calculations that are only capable of being implemented on computers, either algebraically or by means of numerical analysis implemented using interval arithmetic. The latest example is Hales' proof of the Kepler conjecture. As computers grow in power and become steadily more indispensable for the completion of proofs, our awareness of our mental limitations is bound to increase. Philosophers will have to rethink their theory of knowledge to take into account the increasing amount of important technical knowledge that depends on a symbiosis between our computers and ourselves. We will have to accept that even in mathematics there are facts that we have to take on trust, because our computers tell us that they are so. The exceptionalism of human reason will be replaced by the realization that we have invented a method for resolving some types of problem, but may not always fully understand its application.

We used to tell our students not to believe any theorem that we told them unless they had checked every line of the proof themselves and were personally convinced of its correctness, but this is no longer possible. The classification of finite simple groups is a body of work so large that no individual can be said to fully understand it. Those in the subject say that the Classification Theorem is a theorem "by the consent of the community", a concept that is foreign to some of us. This is in no way a criticism of a magnificent achievement, but it again suggests that the nature of mathematics is changing. How does one decide what constitutes a fundamental change in a principled way?

I suppose that I should hope that machines do not render mathematicians redundant, but actually I do not worry much about it, because I am sure that it will not happen in my lifetime. Roger Penrose has argued that it is in principle impossible, for reasons based on Gödel's theorem, but many people find his arguments wholly unconvincing. His discussions of the limitations of computers are lengthy, and not without merit, but they do not take account of the possibility that fundamentally different hardware or software may lead to computers that mimic our thought processes much more closely, at the cost, perhaps, of occasional fallibility. One of the many flaws in his analysis is that he assumes without discussion that the reasoning powers of human beings have no such limitations. Of course he is a Platonist, so he can simply appeal to his direct perception of the Platonic realm, but the only evidence he provides for this possibility is his intuition.

What are the most important open problems in the philosophy of mathematics and what are the prospects for progress?

The more I think about it the more I feel that some philosophers of mathematics have dug themselves into an unproductive hole. My comments below are inevitably generalizations, and do not apply to all philosophers (how could they?) but they do describe a significant part of what is currently being written. The first problem is that for many philosophers mathematics means pure mathematics, with applied mathematics and its scientific applications regarded as being of lesser importance. The second problem is the depersonalization of the subject, which is frequently regarded not as an activity of mathematicians but as an absolute entity, to be studied without reference to its history or applications to the natural world. This is harmless as far as mathematicians are concerned, but it makes a philosophical analysis of the subject more or less impossible.

Those who try to justify the exceptional status of mathematics by reference to its crucial importance for our most fundamental scientific theories often fall into the same Platonic trap. Almost everyone identifies the most fundamental scientific theories as being those that are the most mathematical and conform most fully to the reductionist model of science. Anyone who attempts to question reductionism or who suggests that science is a human activity in which we construct models of the world that help us to understand it risks being labeled a cultural relativist. Scientific knowledge is supposed to be perfectible, and the final formulation is assumed not to involve human beings in any way.

It is puzzling that people are prepared to base any philosophical arguments about the status of mathematics on the currently most admired scientific theories. Certainly the three most prominent theories are astonishingly accurate and have very wide ranges of applicability, but it is *philosophically* disturbing that they involve totally different ontologies. In Newtonian mechanics space and time are quite different entities and particles obey dynamical laws involving gravitational forces that act instantaneously at a distance. In general relativity gravity is not a force but is manifested by the curvature of space-time, and the technology is that of differential geometry. Quantum theory is as different from both as anything could be, and its ontology would be rejected as ludicrous if it were not so successful at predicting phenomena at the atomic level. (Actually there are a number of different ontologies, all bizarre, and all regarded by their advocates as obviously correct.) Finally string theory, which really hardly exists yet, supposes that the world is ten-dimensional. There is no clear evidence that space is continuous at a small enough scale, and the fact that our best theories treat it that way may be precisely what is wrong with them.

It is unfortunate that string theorists talk about the world in Platonic terms, as if it can be identified with a set of mathematical equations. In some cases this may be simply for economy of language, but it is clear that others consider that their Theory of Everything (TofE) will be the final and whole truth, as its unfortunate name suggests. Another problematical aspect of string theory is its current lack of any real relationship with experimental physics. As a *mathematical* enterprise it is a very worthy goal – having two well-confirmed but fundamentally inconsistent theories, quantum mechanics and general relativity, both of which generalize Newtonian mechanics, is a highly unsatisfactory state of affairs. We expect that the effort to construct a TofE will eventually be successful, and this will be its own reward, even if it leads to no new physics. Little remains of early optimism that there would prove to be only one such theory and that it would permit the computation of the fundamental constants of nature. What the TofE will not do is herald the end of physics. Indeed the best that we can probably hope for is a better understanding of the properties of the exotic fundamental particles observed in accelerators.

Three people stand out as resisting the pressure to depersonalize mathematics. Lakatos emphasized that mathematics can only be understood as the activity of human beings, and that the process by which theorems emerge into the world has no connection with the way in which the subject is normally presented in journals. Every mathematician recognizes this fact, which explains their preference for learning the subject via discussions at a blackboard rather than by reading journals. If philosophers dismiss this process as belonging to the sociology of mathematics, then they are left with an impoverished subject that has only a remote connection with what mathematicians do, or with the history of the subject.

The second person worth mentioning as resisting the reductionist/Platonist orthodoxy in science is Popper. He emphasized that scientific theories were human creations, and that we can never prove their correctness. In later life he also argued that human social constructs, entities in his World 3, were as real as physical objects, because the former can affect the latter. Such a viewpoint does not involve rejecting the astonishing successes of the reductionist method over the last four centuries. As a *methodology* reductionism has been extremely successful in the physical sciences, but that does not imply that it is equally appropriate in all other fields of science. In ecology and other biological sciences multiple points of view, including those that are teleological in character, are needed to obtain a full understanding. It would, in particular, be difficult to give a meaningful explanation of organ transplantation that did not mention its purpose.

Popper was strongly influenced by Kant, who described mathematical knowledge in terms that contrast strongly with those of Plato. Unfortunately Kant's style of writing is almost impenetrable, and his detailed comments about Euclidean geometry were wholly misguided. Kant accepts the reality of the external world while identifying our theories about it as creations heavily conditioned by our own mental natures. He identifies space and time, or geometry and arithmetic, as being particularly important. His description of mathematics as being synthetic a priori is perhaps even more appropriate today than it was at the time. The a priori character of the subject is based upon its dependence on logical argument. He breaks decisively with Plato by holding that the synthetic aspect of mathematics results from our inevitable use of concepts based on innate characteristics of the human mind. For Kant our appreciation of the external world has to start from these concepts, whatever the 'world in itself' might be like. He is right in the sense that experimental psychology confirms that we have inborn expectations about the world, and interpret it in the light of these. The extent of these inborn expectations and the degree to which they are modified by later experience is the subject of active research, but their existence is undeniable. For vision they are manifested by the wiring of the retina and visual cortex, which develop along genetically programmed lines, and include specific circuits for detecting edges and their orientations.

Mathematical exceptionalism has a long history. As far as we know it started with the Pythagorean mystery religion, which no doubt influenced Plato. Today it satisfies the desperate need of Western civilizations to find suitable prophets for their age, the more mystical and incomprehensible the better. Some mathematicians and scientists, probably unknowingly, have found that they acquire fame in proportion to their vagueness by making sweeping claims about the future course of scientific understanding. Those who question their claims are dismissed as lacking the deep insights of the masters, but unfortunately high creativity in one field is no guarantee of even basic common sense in another.

The Platonic fallacy has been with us for over two thousand years, but its only function has been to make mathematicians feel comfortable about their subject. It has not solved any philosophical problems about mathematics, and has prevented many from being addressed sensibly. It is probably too much to hope that it will finally be laid to rest in this century, but I nevertheless hope that it may eventually disappear. We do not need quasi-religious mysticism in philosophy, and should not tolerate it simply because earlier generations have done so.

9

Michael Detlefsen

Professor of Philosophy
University of Notre Dame, USA

Why were you initially drawn to the foundations of mathematics and/or the philosophy of mathematics?

In a way, my interest in mathematics and its foundations began my first year in high school in a geometry course. I enjoyed working on geometrical problems and was interested in what the restriction to compass and straightedge really entailed. Sometime during the course, our teacher told us about the classical insolubilia. I found the notion of an unsolvable problem curious, though, and wondered why you couldn't solve the trisection problem, say, by allowing use not only of compass and straightedge but other instruments.

When I asked my teacher about this, she gave a straightforward, practical answer that I found persuasive. The point of restricting the means available to solve problem was to force us to think harder and, so, to better improve our minds. Then she gave a couple of examples. One was the construction of equilateral triangles with sides of a given length, the other the trisection problem.

Why did we go to so much trouble to show how to construct equilateral triangles, she asked. Why didn't we just use our rulers and protractors and be done with it? The answer was that we wanted to strengthen our capacities as reasoners not as measurers. Effective use of rulers and protractors, she said, was mainly a matter of skill in measuring. It didn't increase our skills as reasoners in any discernible way. Restricting ourselves to compass and straightedge constructions, on the other hand, forced us to think more subtly and creatively and this lead to a general strengthening of our capacities as reasoners.

This seemed sensible to me, and I've since learned that it's one of the most common defenses given throughout history for the study of mathematics (see the remark from Bacon quoted in item 2 below). I found myself wondering, though, whether there might be ways to make unsolvable problems solvable without trivializing them or making them worthless for purposes of intellectual development. My teacher told me that there were and she pointed me to some reading concerning these matters. Solvability and unsolvability have been a major interests of mine ever since.

In my first three years in college I studied mainly mathematics and physics. In my third year, at the urging of friends, I signed up for a certain philosophy course. Among the readings were portions of Plato's Republic, most of which I found confusing. I was particularly puzzled by the passage in Book VII where Socrates criticized the geometers of his day for their preoccupation with constructions.

This made no sense to me. The things Socrates seemed to regard as distractions were in my view the very heart of geometry. I thus suspected that there was something weird about the constructions of the geometers Plato was criticizing. With this in mind, I made an appointment with my professor to see if he could tell me more. He surprised me by saying that it was the classical geometrical constructions that Plato was criticizing. His criticism was of the use of construction generally, not of some particular types of constructions. As he developed Plato's thinking for me, I became impressed with how subtle and interesting the issues were and wanted to understand them better. My interest continued to increase and I decided to give graduate studies in philosophy a try.

What examples from your work (or the work of others) illustrate the use of mathematics for philosophy?

It depends on what's meant by "use of mathematics." Earlier philosophers and mathematicians often remarked the "intervenient" usefulness of (studying) mathematics. Bacon (*The Advancement of Learning* (1605), Bk. 2, VIII, 2), for instance, wrote:

> In the Mathematics I can report no deficience, except it be that men do not sufficiently understand the excellent use of the Pure Mathematics, in that they do remedy and cure many defects in the wit and faculties intellectual. For if the wit be too dull, they sharpen it; if too wandering, they fix it; if too inherent in the sense, they abstract it. So that as tennis is a game of

no use in itself, but of great use in respect it maketh a quick eye and a body ready to put itself into all postures; so in the Mathematics, that use which is collateral and intervenient is no less worthy than that which is principal and intended.

This seems right to me, so I'd certainly say that mathematics has had an "intervenient" effect on my work. Since this is true for virtually everyone, though, I doubt it's what you had in mind. I'll thus briefly discuss a couple of examples that may be closer to what you had in mind.

Let me begin by saying, though, that it's important for the philosophy of mathematics to consider not only influences of mathematics on philosophy but influences of philosophy on mathematics. History offers at least as many examples of the latter as of the former, and they seem generally to be as important for the philosophy of mathematics. This noted, I'll now describe one example of each type.

Methods of Discovery and Methods of Demonstration

The first concerns an old distinction between two types of methods that has been taken to apply to careful, systematic forms of inquiry generally. Cicero emphasized this distinction, terming methods of the one type methods of "discovery" or "invention" and methods of the other type methods of "adjudication" or "demonstration". In time this became the traditional distinction between arts of invention or discovery (*artis inveniendi*) and arts of adjudication or demonstration (*artis iudicandi* or *demonstrandi*).

The basic idea was that certain methods may be adequate for purposes of making a presumptive case for something even though they may not, on final reflection, be adequate for its ultimate disposition. This was an idea that appealed to the practitioners of a variety of disciplines, including mathematics and philosophy, but also the natural sciences and, perhaps most notably, jurisprudence. It was also reflected in the later distinction, so prominent in Kant's thinking (and eighteenth century thinking generally), between *discursive* and *intuitive* reasoning.

Let's begin by considering the form this distinction has taken in western jurisprudence. We see it in the principle of *corpus delicti*, which requires evidence not only for justified *conviction* but for justified *trial* as well. To justify the undertaking of a trial, two types of evidence must be produced—first, evidence of the "existence" of a crime to be tried, and secondly, evidence of the identity of the perpetrator. In the case of murder, this has typically meant evidence both of the death of a person as the result of an act, and evidence of the criminal agency of an identified person as the means of that death. (Other types of crimes call for other types of evidence of course. But all legitimate trial requires evidence both of the existence of a crime and of the identity of the alleged criminal(s).)

In addition to this, there is another requirement, namely, that the evidence produced should not be merely *discursive* in nature, the type of evidence that a coherent story or narrative may in and of itself provide. Rather, it is to have a substantial material component, as, for example, in the case of a trial for murder, the production of a corpse as evidence that a death has occurred.

This general scheme of distinctions has long been recognized by both philosophers and mathematicians. An early statement can be found in Proclus' commentary on the *Elements* as part of a defense of Euclid's ordering of the first four propositions of Book I. Proclus sees the first three propositions as wanted for the establishment of the existence of equilateral triangles, and he sees this as being in turn a prerequisite of rightly undertaking to prove the fourth proposition, which asserts an essential property of equilateral triangles. He thus wrote:

> For unless he had previously shown the existence of triangles and their mode of construction, how could he discourse about their essential properties? Suppose someone ... should say: "If two triangles have this attribute, they will necessarily also have that." Would it not be easy ... to meet this assertion with "Do we know whether a triangle can be constructed at all?" ... It is to forestall such objections that the author of the *Elements* has given us the construction of triangles ... " (*A Commentary on the First Book of Euclid's Elements*, G. R. Morrow (trans.), pp. 182–183)

Leibniz adopted a similar view, claiming that constructive or genetic definitions were valuable as means of showing the reality (or at least the possibility) of the items defined, and that such evidence was a general prerequisite of proper demonstration.

> ... the concept of circle put forward by Euclid – namely, that it is the figure described by the motion of a straight line in a plane about one fixed end – affords a real definition, for it is clear that such a figure is possible.

> It is useful ... to have [such] definitions ... beforehand. ... [for] we cannot safely devise demonstrations about any concept, unless we know that it is possible; for of what is impossible ... contradictories can ... be demonstrated. ("Of Universal Analysis and Synthesis; or, of the Art of Discovery and of Judgement" (1683), pp. 12–13 in *Philosophical Writings [of] Leibniz*, ed. and tr. Mary Morris and G.H.R. Parkinson).

At various times in its history, then, mathematics has recognized the legitimacy of two broadly different types of justificatory enterprises, those in which propositions are demonstrated, and those in which they're marked as reasonable targets of demonstration. This recognition seems to have reflected the acceptance of similar divisions of evidensory type outside mathematics.

Nor is this merely the conviction of bygone eras. Many mathematicians of the nineteenth and twentieth centuries (most notably, perhaps, Hilbert) retained the same idea in the form of a division between *real* and *ideal* methods. And very recently Arthur Jaffe and Frank Quinn ("Theoretical mathematics": Towards a cultural synthesis of mathematics and theoretical physics", *Bulletin of the American Mathematical Society* 29 (1993): 1–13) have urged a new form of the distinction. They advocate the use of freer, more efficient "theoretical" methods (so called because they resemble the methods used in theoretical physics) to generate initial hypotheses and to outline justifications. These hypotheses and justifications are then to be converted into rigorous reasoning by mathematicians particularly skilled in such work.

Jaffe and Quinn see the role of rigorous proof in mathematics as "functionally analogous to the role of experiment in the natural sciences" (p. 2), and they therefore propose to divide two types of mathematical research—a more intuitive or speculative "theoretical" type aimed at efficient discovery, and a more rigorous, conventional type aimed at confirmation.

Full practical adoption of this division, they suggest, should hasten the advancement of mathematics in much the same way that it hastened the advancement of modern physics. They thus propose that it be institutionalized, both in the training of mathematicians and in the longer term evaluation of their work. The distinction between methods of discovery and methods of demonstration in mathematics is, it seems, still with us today.

9. Michael Detlefsen

Symbolic Algebra and the Development of Formalism

The second example I want to discuss concerns an influence that ran in the opposite direction, from mathematics to philosophy. Its inspiration was the successful use of symbolic methods in algebra in the seventeenth century, a use which challenged the way philosophers had traditionally conceived of reasoning.

The new conception, advanced principally by Leibniz and Berkeley in the eighteenth century, was not universally endorsed. It did, however, in time become influential. Its cornerstone was the non-traditional view that language gains cognitive significance not only through its capacity for semantic interpretation but also through its ability to function as a calculus.

Berkeley thus asserted a broadly instrumental function of language, and he took as his prime evidence the symbolic methods of the algebraists.

> ... it is a received opinion, that language has no other end but the communicating [of] our ideas, and that every significant name stands for an idea. ... [But] a little attention will discover, that it is not necessary (even in the strictest reasonings) [that] significant names which stand for ideas should, every time they are used, excite in the understanding the ideas they are made to stand for: in reading and discoursing, names being for the most part used as letters are in *algebra*, in which though a particular quantity be marked by each letter, yet to proceed right it is not requisite that in every step each letter suggest to your thoughts, that particular quantity it was appointed to stand for. ... the communicating of ideas marked by words is not the chief and only end of language, as is commonly supposed. (*A Treatise concerning The Principles of Human Knowledge*, 37)

The end of language was thus not, in Berkeley's view, "merely, or principally, or always, the imparting or acquiring of ideas" (*Alciphron: or the Minute Philosopher*, 307). Rather, it was sometimes something more instrumental in character, something which allowed it to support the pursuit of knowledge even when it made use of signs for which there is "no possibility of offering or exhibiting any ... idea to the mind: for instance, the algebraic mark, which denotes the root of a negative square, hath its use in logistic operations, although it be impossible to form an idea of any such quantity" (loc. cit.).

Berkeley's vews of language and reasoning were thus inspired by the success of symbolic methods in mathematics. In time these ideas developed into a view that saw the syntactical use of language as reflecting (or perhaps even constituting) the laws of human reasoning. Boole held a view of this sort. As he put it, "language is an instrument of human reason, and not merely a medium for the expression of thought" (*An Investigation of The Laws of Thought*, 24). In addition, the operations of language express "operations of the human intellect" (loc. cit.) in such a way that "in studying the laws of signs, we are in effect studying the manifested laws of reasoning" (loc. cit.).

Belief in this correspondence between the operations of the human intellect and syntactical operations on language then turned again to inspire developments in mathematics. Prominent among these was Hilbert's proof theory, the goal of which was described as the making of "a protocol of the rules according to which our thinking actually proceeds." ("Foundations of Mathematics" (1928), 475). He did this in the belief that "[t]hinking ... parallels speaking and writing: we form statements and place them one behind another" (loc. cit.) and that even in "every day life one uses methods and concept-formations which require a high degree of abstraction and which only become intelligible by means of an unconscious application of the axiomatic method" ("Naturerkennen und Logik", 380). Hilbert found the investigation of this phenomenon – the parallel between syntactic manipulation and reasoning – so compelling that he made it the object of his proof theory.

There are many other examples that one might give of significant interactions between mathematics and philosophy. I mention the two above because they've had an important bearing on my own attempts to understand a central phenomenon of mathematics—namely, its willingness to use ideal or imaginary elements.

What is the proper role of philosophy of mathematics in relation to logic, foundations of mathematics, the traditional core areas of mathematics, and science?

The philosophy of mathematics has, I think, gained considerable from developments in mathematical logic. An example of this is the way in which developments in proof theory have given us a more precise understanding of the conditions required for the success of Hilbert's Program. I think good philosophy of mathematics has something to offer proof theory too. Here too, I think, the proper evaluation of Hilbert's program is a case in point.

Hilbert prized ideal reasoning for its efficiency or convenience, however exactly he might have conceived that. In principle, he was committed to defending the use of only so much ideal reasoning as is gainful, and that is ideal reasoning that improves upon the efficiency of real reasoning. Let's say that an ideal proof of a real theorem **r gainful** if is more efficient/convenient than any real proof of **r**. And let **I_g** be the set of theorems that results from the removal of all real theorems of **I** for which there is no gainful proof in **I**. The Hilbertian is in principle committed not to establishing the consistency of **I** but only the consistency of (an appropriate axiomatization of) **I_g**. It thus behoves us to try to identify, and thence to investigate, the gainful subsystems of ideal systems to determine which if any may admit of the type of consistency proofs (specifically, finitary consistency proofs) that Hilbert's program required.

It's possible, of course, that **I_g** will *not* be a proper subsystem of **I**, either generally or in any significant particular case (e.g. where **I** is $\mathbf{PA}^2$ or **ZFC**). Moreover, even if there are important cases of **I** where **I_g** *is* (axiomatizable as) a proper subsystem of **I**, the **I_g** in those cases may still subsume the whole of finitary reasoning. I think our knowledge of these matters is currently quite weak, though, and, for the most part, even unexplored. This seems important because the ultimate disposition of Hilbert's Program depends upon our having such knowledge. To be a little more exact, it depends on our having answers to questions such as the following.

What notion of complexity is in play here? Is it a form of complexity that measures the difficulty of determining of a given object that it is a proof of a certain type? Or ought it rather to be a "discovermental" notion of efficiency—that is, a notion of complexity that measures the difficulty of discovering a proof in the first place? Did Hilbert think ideal methods made verification of proofs easier? Or did he think they made them easier to discover? What he wrote suggests the latter, but proof theorists have mainly paid attention to the former type of complexity. It seems to me that a properly careful philosophical treatment of Hilbert's Program brings this to light. Hence my claim that not only does logic have something to offer to the philosophy of mathematics, but philosophy of mathematics has something to offer to logic too. It can help it to see what it doesn't know and what may be important to know. For example, let T be a formal theory that contains ideal proofs and to which Gödel's theorems apply, and let T_g be that subsystem of T which contains all and only those real theorems of T for which no finitary proof is as discovermentally efficient as is some ideal proof in T. It seems to me that we don't generally know the answers to questions of the form: Does Gödel's Second Theorem apply to T_g? Nor do we have a precise understanding of how to define or measure discovermental efficiency.

As regards the relation of the philosophy of mathematics to the traditional core areas of mathematics, I'd say that in their actual historical development, they form part of the data for the philosophy of mathematics. Similarly for applications of mathematics to the empirical sciences.

What do you consider the most neglected topics and/or contributions in late 20th century philosophy of mathematics?

Too little attention has, I think, been given to the differences that separate the different areas of mathematics and the different types of reasoning that appear with regularity in the history of mathematics. Correspondingly, too much attention has been given to developing grand schemes that seek a single "essence" for the whole of mathematics. It may be that this has been encouraged by the development of powerful conceptuo-reductive systems in the foundations of mathematics (e.g. set theory and category theory) and the development within mathematics of preferences for one type of reasoning (e.g., what some have called "algebraic" reasoning) over all others.

Without denying the value of some of these developments, they may nonetheless have undesirable consequences for the philosophy of mathematics. They may, in particular, cause it to underestimate the variety and complexity of the phenomena for whose explanation it is ultimately responsible.

The division of methods (methods of discovery vs. methods of demonstration) discussed in my response to question 2 is a case in point. The attempt to develop efficient sub-demonstrative methods of reasoning is, I think, a persistent feature of historical mathematical practice. This persistence is not, however, well-captured by monolithic views that attempt to see all of mathematics as serving the same end (e.g., the description of structure). Such views may help us to see saliencies amidst the details of mathematical history. They may also, though, encourage its oversimplification.

To be more specific, I'd say that one neglected topic of later 20th century philosophy of mathematics is the persistent tendency throughout the history of mathematics to introduce ideal or imaginary elements.

A second neglected topic is that concerning objectivity in mathematics. What might reasonably be called the classical view of objectivity sees the distinction between objective and subjective methods as centering on the extent to which the methods in question are within our power to control. Genetic or constructive definitions and proofs are treated as superior in objectivity because they're taken to give evidence that's "hard" in the sense of going beyond the production of merely coherent stories spun by individual minds.

In modern times (beginning especially with the seventeenth century), an alternative conception of objectivity – what I'll call the "invariance" conception – became increasingly important. On this conception, the basic idea is that objective science is the study of those properties of the objects studied that are "in" those objects themselves, as distinct from being imposed on our thinking about them by conditions of our own cognition (at least not conditions of our *individual* cognition).

This was the type of concern that philosophers sometimes expressed through the adoption of a distinction between the primary and secondary qualities of things. Primary qualities of objects were seen as properties of objects that were in the objects themselves and could not be removed from them. In particular, they could not be separated from them by variation of perspective on the part of an observer.

Secondary qualities, on the other hand, were taken to be just the opposite, that is as separable from the objects by sufficient variation of perspective.

Something like this conception of objectivity seems to have played a role in the thinking both of early algebraists, the projective geometers of the eighteenth and nineteenth centuries and various twentieth century figures. Wallis, for example, defended his algebraic approach to geometry because he believed it offered a more satisfying treatment of the invariant figures of geometrical objects (e.g. quadrature laws for plane figures) than did the methods of classical geometry (*Treatise of Algebra* (1685), 291).

> ... beside the supposed construction of a Line or Fig-

ure, there is somewhat in the nature of it so constructed, which may be abstractly considered from such construction; and which doth accompany it though otherwise constructed than is supposed. As for instance, a Circle (according to *Euclid's* construction,) is such a Figure as may be described by carrying about of a Straight Line (till it return thither from whence it began,) whose one end remains fixed at (what we call) the Center, and the other describes (in the same Plain) a Curve. But the same may be described also (as *Apollonius* shews us) by cutting a Cone by a Plain Parallel to the Base; or (as *Serenus* shews) by such cutting a *Cylinder* parallel to the Base; or (as others also shew) by cutting a Sphere by any Plain however situated. Yet are all these Circles (however constructed) of the same nature, and have the same properties appertaining to them.

Similar statements may be found in the writings of the projective geometers (cf. the introduction to Poncelet's *Traité des propriétés projectives de figures* ...) and in the writings of philosophers of mathematics of the twentieth century. Waismann, for example, expressed the idea this way (*Introduction to Mathematical Thinking*, 179–81):

> On reconsidering the various geometries, we find that the spatial transformations which produce a definite class of configurations always exactly form a group. Thus, the transformations which leave the elementary-geometric properties of a structure unchanged are the motions, the similarity transformations and the process of reflection. ... a group which is designated as the *principal group* of the spatial transformations. ... If the principal group is replaced by a more comprehensive one, then only some of the properties remain invariant. ... The more comprehensive the group is ... the more we penetrate into the depths. The properties which metric geometry investigates are the easiest to destroy; they lie, so to speak, at the uppermost stratum. The affine properties lie somewhat deeper; the projective, still deeper.
>
> ... On going still further to the most general one-to-one point transformations, we arrive at the standpoint of

the *theory of sets*. From this point of view the distinction of dimensions is also vague: a line segment, a surface, a solid can be mapped into one another point for point, that is, they are only different representatives of one class. Of the whole rich world of geometrical forms there now remain only very few traits, such as the distinction between finite and infinite point sets.

There are thus at least two different conceptions of objectivity that have figured importantly in the history of mathematics. Moreover, since they suggest different ideals for proof, there seem to be consequences of adopting one rather than the other. On the classical view, the ideal method of proof is one that features construction. On the invariantist view, it's the method of maximal invariance.

Which ideal of objectivity one adopts will play a role in determining which methods of proof are seen as best. On what basis does one decide? And are the two ideals equals, or is one of them superior to the other (when both are possible)? These are questions that seem to have a potential bearing on mathematical practice and they are, I believe, neglected questions. The problem concerning the nature of necessity/possibility in mathematics mentioned in the next response also belongs to the category of neglected questions.

What are the most important open problems in the philosophy of mathematics and what are the prospects for progress?

I'd include the problems noted in various of the responses already given (e.g. what is the place of different types of methods in mathematics and how ought we to understand objectivity in mathematics). Of others that might be mentioned, I'll restrict myself to one concerning the nature and extent of "freedom" in mathematical theorizing or, relatedly, the nature of necessity/possibility in mathematics. I'll begin by recalling a remark of Gauss that raises some of the central issues.

Gauss once criticized certain mathematicians of his time (e.g., Euler, Lagrange and D'Alembert, to name but three) for referring to imaginary and complex numbers as "impossible" numbers. Gauss thought this to be bad terminology because it encouraged the false view that invention in mathematics is something like creation from whole cloth. He thus wrote (*Demonstratio nova theore-*

matis omnem functionem algebraicam rationalem integram unius variabilis in factores reales primi vel secundi gradus resolvi posse, 1799, reprinted in *Werke Sammlung*, vol. 3, p. 6):

> If someone would say a rectilinear equilateral right triangle is impossible, there will be nobody to disagree. But, if he intended to consider such an impossible triangle as a new species of triangle and to apply to it other qualities of triangles, would anyone refrain from laughing? That would be playing with words, or, rather, misusing them.

Gauss' point, at least in part, seems to have been that imaginary numbers, and mathematical inventions generally, are not like rectilinear, equilateral right triangles. But what, exactly, did he see as the differences? And what did he take to be so wrong with rectilinear equilateral right triangles?

It may be tempting to answer that the notion of a rectilinear equilateral right triangle violates one or more of the basic laws governing the notions of right angle and equilateral, rectilinear triangle.

Fair enough, but isn't it also true to say that the notion of a square negative magnitude violates one or more of the fundamental laws concerning the notion of magnitude? The complex numbers do not, after all, satisfy both the following laws, each of which would seem to be pretty basic to the concept of magnitude or quantity.

L1 For every x, either $x < 0$ or $x = 0$ or $x > 0$.

L2 For every x, y, if $x > 0$ and $y > 0$, then $x + y > 0$ and $xy > 0$.

This raises what is to my mind an important question for the philosophy of mathematics—namely, what the conditions of genuine mathematical possibility/freedom are and whether and why they allow such things as the introduction of imaginary numbers but not the introduction of equilateral right triangles.

10
Solomon Feferman

Patrick Suppes Family Professor of Humanities and Sciences, Em.
Professor of Mathematics and Philosophy, Em.
Stanford University, USA

Why were you initially drawn to the foundations of mathematics and/or the philosophy of mathematics?

I'm a philosopher by temperament but not by training, and a philosopher of logic and mathematics in part, as I shall relate, by accidents of study and career. Yet, it seems to me that if I was destined for anything it was to be a logician primarily motivated by philosophical concerns.

When I was a teenager growing up in Los Angeles in the early 1940s, my dream was to become a mathematical physicist: I was fascinated by the ideas of relativity theory and quantum mechanics, and I read popular expositions which, in those days, besides Einstein's *The Meaning of Relativity*, was limited to books by the likes of Arthur S. Eddington and James Jeans. I breezed through the high-school mathematics courses (calculus was not then on offer, and my teachers barely understood it), but did less well in physics, which I should have taken as a reality check. On the philosophical side I read a mixed bag of Bertrand Russell, John Dewey and Alfred Korzybski (the missionary for "General Semantics" in *Science and Sanity*, a mish-mash of the theory of types, non-Aristotelian logic and colloidal chemistry, among other things). Also, I was fascinated by, and bashed my head against, Rudolf Carnap's *Logische Aufbau der Welt*, but couldn't penetrate it. Still, I should have taken its attraction for me as another sign. One thing I did know for sure, and that was that I wanted to have an academic career and become a professor. What the source of that was is a bit of a mystery to me, since my parents and their friends were working class and I had no personal role models. But I suppose I learned in one way or another that that was the way to go if I were to devote myself to theoretical research.

In 1945, I applied to both UCLA and CalTech for mid-year entry to undergraduate studies. I leaned toward UCLA since it was practically free, it had a broader curriculum, it was co-ed, and many of my friends were going there. By contrast, CalTech was focussed on science and engineering, it had an all male student body in those days (and did not become co-ed until many years later), and none of my buddies applied there. It was also more prestigious. Perhaps I would have had a different career path if I had not passed the entrance exam for CalTech and still had physics so strongly on my mind. Tuition expense was a serious obstacle, but I was offered a part-scholarship and a job as a lunch-time waiter in the *Athenaeum*, CalTech's faculty dining hall, and my proud parents made ends meet somehow or other. (A high point one day as a student waiter was serving a boiled egg and salad to Robert Oppenheimer, whom I met again years later in his capacity as Director of the Institute for Advanced Study in Princeton, when I was a fellow there in 1959–60.)

In my courses during the first two years at Caltech, mathematics was as before a breeze and fun, while in my physics courses I found that I lacked even minimal physical intuition. Still, mathematical physics was my goal, and for that the book that was touted for students was Harry Bateman's *Partial Differential Equations of Mathematical Physics*. Looking through it made clear to me that that wasn't at all the kind of thing I was after. So, in my junior year, as a kind of fall-back position, I switched majors to mathematics. But there I found that I had to enter a new mind-set, that of pure mathematics and theorems to be proved rather than problems to be solved and techniques to be mastered. Of all the courses that I took from then on, only one appealed as a possible direction for further study, and that was an introduction to logic taught by Eric Temple Bell—known to mathematicians as a number-theorist and author of the romantic and historically flawed, *Men of Mathematics* and *The Queen of Sciences*, and to science-fiction aficionados of the day through a pseudonym, John Taine (*Green Fire, The Iron Star*, etc.). The course, which hardly got beyond propositional calculi of various kinds, was a hodge-podge because Bell did not really know anything substantive in logic; I learned years later that he had a fatal attraction to Lukasiewicz' three-valued logic (in his *The Search for Truth*). Despite its incoherent presentation, the material of that course resonated with me, but there was no follow-up to be had at CalTech.

For a career in academia, it was clear I would have to go on to graduate work in mathematics, and in 1948 I applied to UC Berkeley and the University of Chicago. I was accepted at both places, but only Berkeley came through with an offer of a teaching assistantship, so that pretty much clinched it. (Also I had a strong personal reason to prefer Berkeley—a romantic interest.) In my first year, I took some of the basic required courses in algebra and analysis, and I met Fred Thompson, who was working on a PhD with Alfred Tarski. Thompson idolized Tarski, raved about him no end, and urged me to take his course in metamathematics, which I did the following year. To quote myself from the biography that I co-authored with Anita Burdman Feferman, *Alfred Tarski. Life and Logic* (2004), "I knew immediately that this was to be my subject and Tarski would be my professor. He explained everything with such passion and, at the same time, with such amazing precision and clarity, spelling out the details with obvious pleasure and excitement as if they were as new to him as they were to us." In the following years I went on to take courses in model theory, set theory and universal algebra with Tarski and became a regular attendee at his seminars, which in those years concentrated on algebraic logic. My introduction to recursion theory and Gödel's incompleteness theorems came via Andrzej Mostowski's 1952 book, *Sentences Undecidable in Formalized Arithmetic*, through its use as the text of a course taught by Jan Kalicki (a promising young logician who, tragically, died in an automobile accident in the fall of 1953). There were no courses in proof theory.

I did indeed end up working toward a PhD with Tarski, but the excellent initial progress I had made in my studies with him did not presage the difficulties that I would have in arriving at a dissertation result to his satisfaction. (I was not the only one with this problem.) That story has been told in our biography of Tarski, and in more detail in my article "My route to arithmetization" (1997). Briefly, Tarski suggested two problems for me to solve, one on cylindric algebras, and the other on a decision procedure for the ordinals under addition, both of which I attacked dutifully but with only partial success. Fate intervened when I was drafted into the US Army in the fall of 1953 (fortunately not a time of active war for the US, since it was post-Korea and pre-Vietnam). After basic training, I was assigned to a unit in the Signal Corps at Fort Monmouth, New Jersey, doing research on kill probabilities of hypothetical missile attacks on major cities and target sites in the US. My thoughts about this alternated between bemusement with the essential unreality of our calculations and anxiety about the possible reality of the scenarios with which they dealt. In what leisure time I had during off-hours, I read and reread Kleene's *Introduction to Metamathematics*, and that added significantly to my understanding of recursion theory and Gödel's theorems, as well as oriented me toward Hilbert's finitist consistency program. To my surprise during that period, Alonzo Church, as the *Reviews* editor of *The Journal of Symbolic Logic*, asked me to take on an article by Hao Wang. That concerned an arithmetized version of Gödel's completeness theorem that extended an earlier version in vol. II of Hilbert and Bernays' *Grundlagen der Mathematik*. (I think Dana Scott, who was by then studying in Princeton, suggested my name to Church.) That led me to the question as to how, precisely, one should deal with formalized consistency statements in general, and thence directly into my work on the arithmetization of metamathematics. When I was released from active army duty and returned to Berkeley in 1955 I proposed that to be the new topic of my dissertation.

As it happened, Tarski was on sabbatical leave in Europe that year, and Leon Henkin agreed to help supervise my work in his absence. With his constant encouragement in the following months I obtained a number of good new results and I sought Tarski's approval to have them form the main part of my thesis. Because it was out of the mainstream of his interests, and perhaps because it dealt with problems arising from Gödel's work rather than his own, he was initially resistant to that. But he consulted Mostowski and then acceded, though still with some reservations, after he received the latter's quite positive report.

Crucial to me in that period, and, as it turned out, for many years following, was my contact with Georg Kreisel. To quote myself again, in "My route to arithmetization" I wrote: "I first met Kreisel during the period in early 1956 when I was well into the research for my hoped-for dissertation; Kreisel happened to be visiting Berkeley for a month or so at that time. Our initial personal contact was magical for me: I had hardly to begin explaining what I had done and what I was in the process of working on, to see that Kreisel understood immediately, and that it related to things he had thought about and to a whole body of literature in which he was completely at home. His positive reception of my ideas confirmed my views of the significance of what I was up to, and added to my determination to make this work my thesis, despite Tarski's reservations ... the boost provided by Kreisel's quick appreciation was psychologically crucial at that agonizing time. In addition, Kreisel opened up a new world to me through his interests in constructivity, predicativity and proof theory, interests which I was naturally attracted to and which would come to dominate my own subsequent work."

I wrote up the dissertation itself during the academic year 1956–57 at Stanford University, where I had been appointed to an instructorship in mathematics and philosophy, and its results were eventually published in 1960 under the title, "Arithmetization of metamathematics in a general settting".

The influence of Tarski and Kreisel was decisive for me, the former in how I carried out my work and the latter in what I worked on. In their own pursuits, both were highly conscious of aims and programmatic development; for Tarski that was largely mathematical while for Kreisel it was primarily philosophical, though Tarski's work in the 30s on conceptual analysis of semantical notions has also been of great philosophical significance. Tarski emphasized clarity and precision of presentation and careful, sequential organization of material; no detail was too small to be overlooked. By contrast, Kreisel emphasized informal rigour and not taking received views for granted; once one had the right ideas, details were supposed to look after themselves. Personally, my relations with Tarski were friendly and frequent throughout the years following my PhD to the time of his death in 1983, but my work was largely disjoint from his, and even where it wasn't he reserved comment. By contrast, Kreisel and I were the closest of colleagues for some fifteen years up to the time we had a rather abrupt and complete falling out in the early 1970s. In any case, stimulating as our contact had been over such a long period, it was time to move on.

What examples from your work (or the work of others) illustrate the use of mathematics for philosophy?

Here are some of the philosophical problems with which I have been concerned off and on over a long period of time. *What is the true reason for incompleteness? How may it be overcome? What ought we to accept once we have accepted given notions and principles? Does mathematics need new axioms? What is the significance of foundational work for mathematical practice?* Inevitably, each has given rise to more specific questions of philosophical relevance that I will also indicate in the following.[1]

To go back to the beginning; the work in my dissertation was driven by the aim to carry out in precise and substantial generality the arithmetization of metamathematics as exemplified in particular by Gödel's second incompleteness (or unprovability of consistency) theorem on the one hand and a formalized version of his completeness theorem on the other. Both involved consistency statements, the latter in the form that a recursively axiomatized theory S is interpretable in Peano Arithmetic (PA) when the consistency of S, Con_S, is added as an axiom. But just what is meant by Con_S in general? That is explained in terms of the arithmetized provability predicate for S, $Prov_S(x)$, and that in turn is determined by an arithmetized definition $Ax_S(x)$ of the set of axioms of S, once we fix the logic to be that of the classical first-order predicate calculus. It turns out that for the formalized completeness theorem it is sufficient for $Ax_S(x)$ to binumerate the axioms of S in PA. But that is not sufficient for the unprovability of consistency theorem, since an example can be given of a binumerative definition of the axioms of PA for which the associated statement Con_{PA} is provable in PA in contrast, of course, to the "canonical" definition. On the other hand, if $Ax_S(x)$ is provably recursively enumerable (r.e.) and S is a consistent extension of PA (and already of much weaker systems) then Con_S is *not* provable in S; any such definition serves to verify the Hilbert-Bernays derivability conditions for $Prov_S(x)$. I showed how one could trade on the difference between these general statements, for example to show that $PA + (\neg Con_{PA})$ is interpretable in PA.

A non-chronological aside: while provably r.e. definitions are sufficient for general formulations of the second incompleteness theorem and other results of the same character in the arithmetization of metamathematics, they are not necessarily *intensionally* correct, so what was still called for was an account of *canonical consistency statements*. As it happens, it was not until the early 1980s that I returned to give full consideration to that matter. My solution, in an improved form in the 1989 paper "Finitary inductively presented logics", was to treat formal systems as they

---

[1] Because of limitations of space, I cannot go here into other parts of my work that I consider to be of philosophical significance, including that on systems of constructive analysis and explicit mathematics, type-free theories of truth, foundations of category theory, relativized Hilbert program, and the limits of logic.

are actually presented to us in practice through the finite inductive generation of various syntactic categories of objects, operations on them, and relations between them; consistency statements are then canonically associated with those. Besides addressing the conceptual issue of finding the "right" framework for general developments, this work was conceived as having potential pedagogical and practical value, the latter via the pursuit of computer implementation of a wide variety of logical systems.

Moving beyond these particular technical and conceptual questions, following my dissertation work I turned to the phenomenon of arithmetical incompleteness itself. What is the reason for it? Can it be overcome? Famously, Gödel in footnote 48a to his 1931 paper on undecidable propositions said that "the true reason for the incompleteness inherent in all formal systems of mathematics is that the formation of ever higher types can be continued into the transfinite ... while in any formal system at most denumerably many of them are available. ... An analogous situation prevails for the axiom system of set theory." To be sure, one can go beyond whatever axioms have already been accepted by adding axioms for the existence of sets that code a truth definition for a model of the previously accepted axioms, and thus prove their consistency. So Gödel's is *one* reason that can be given for incompleteness. But another one that can be given is that it is simply a matter of oversight: whatever has led one to accept a given system S of axioms ought to lead one to accept its consistency Con_S as a new axiom. More generally, one ought to accept an expression of the correctness of S in the form of the *local reflection scheme* $Prov_S(A) \to A$ for each A in the language of S. Moreover, such extensions are formulated without positing the existence of any sets whatever. Unlimited finite iteration of such schemes beginning with, say, PA, still leads to incomplete r.e. systems. So the natural question to ask is, to what extent can arithmetical incompleteness be overcome by the transfinite iteration of consistency statements and more generally of reflection schemata? The first attempt to answer such questions had been carried out by Alan Turing in 1939. He introduced the notion of an *ordinal logic*, which is a uniform means of associating an r.e. system S_a with each $a \in O$, the set of Church-Kleene recursive ordinal notations. Turing showed by an ingenious argument that if one forms the S_a by iterating consistency statements starting with $S_1 = PA$, every true statement A of the form $\forall x R(x)$ with R primitive recursive can be proved in S_a for some $a \in O$ that denotes $\omega + 1$. Turing was particularly interested in obtaining a similar result for statements in the next higher quantificational form, $\forall x \exists y R(x,y)$, via iteration of the local reflection principle; this class includes many interesting open problems in number theory.

In my 1962 paper, "Transfinite recursive progressions of axiomatic theories" (my rechristening of Turing's ordinal logics), I showed that iteration of the local reflection principle is incomplete for $\forall \exists$ statements, but one does obtain completeness for them by iterating instead the *uniform reflection scheme*,

$$\forall x Prov_S(A(num(x))) \to \forall x A(x),$$

i.e. the formalized version of the ω-rule. In fact one obtains much more: for that progression, *every* true arithmetical statement A is provable in S_a for some $a \in O$ at level $\omega^2 + \omega + 1$. Moreover, there are paths P through O and recursive in O, such that every true arithmetical statement A is provable in S_a for some $a \in P$. However, these results, like Turing's, suffer from non-uniqueness: different notations a, b for the same ordinal may have quite different S_a and S_b. Indeed, Spector and I showed in the follow-up paper, "Incompleteness along paths in progressions of theories" (1962), that one has incompleteness with respect to $\forall$-form statements along *any* path P through O that is of the same logical form as O (i.e., one universal set quantifier), and moreover there are many such paths, thus putting us back to square one.

A side conceptual question raised by this work is: What is a *natural system of notations for ordinal numbers*? The paradigm example for that is Cantor's system of notations for the ordinals up to ε_0, the least solution of $\omega^\alpha = \alpha$. The cited completeness results for progressions depend crucially on the construction of non-natural notations that somehow encode the truths to be proved. Ever since Gentzen's proof in 1936 of the consistency of PA by transfinite induction up to ε_0 with respect to a primitive recursive predicate, the question of natural well-orderings has also been of prime significance for proof-theorists in pursuit of Hilbert's consistency program for systems much stronger than PA. In practice, that work applies transfinite induction up to ordinals α much larger than ε_0 for which one has a natural system of notations up to α. Moreover, every such system has a recursive ordering on it and thus is embeddable in an initial segment of O. But simple examples serve to show that recursiveness is far from sufficient as a condition for naturality. This is a problem that has yet to receive a satisfactory answer; some efforts to provide one are surveyed in my lecture text, "Three conceptual problems that bug me" (1996a).

In any case, the main problem with the work on ordinal logics/recursive progressions of theories was the lack of a conceptually motivated restriction on which ordinal notations ought to be accepted. Following Kreisel's own earlier work "Ordinal logics and the characterization of informal concepts of proof" (1960) on a progression related to finitism, he suggested that the choice of notations should be controlled by an *autonomy condition*. That is, one may proceed to an S_a only if it has been proved in some previously accepted S_b with $b < a$ that the ordering of notations up to a is well-ordered, so that a recognizably denotes an ordinal α.[2] On the other hand, the restriction to the language of arithmetic that had been taken before may be considered to be arbitrary. For, once one has accepted a system S as correct, one ought to accept the truth predicate for sentences of the language of S as a new basic notion with its usual closure conditions as new axioms; doing so automatically yields the uniform reflection principle for S. Thus one is led to consider a progression of theories starting with PA obtained by iterating autonomously the adjunction of truth predicates. It turns out that this is equivalent to an *autonomous progression of ramified second-order theories* RA_α, or ramified analysis. Just as Russell ramified the theory of types in order to meet the Vicious Circle Principle and thus satisfy Poincaré's injunction against impredicative definitions, so this could be considered to provide a characterization of the notion of *predicative provability given the natural numbers*. The first question to ask, assuming that, was, what is *the ordinal of predicativity*, i.e. what is the least ordinal not obtained in the autonomous progression of RA_α's? The answer to that was provided in my paper, "Predicative systems of analysis" (1964) and, independently, around the same time by Kurt Schütte. Denoted Γ_0, it is the least ordinal γ not obtained by transfinitely iterating the fixed point process applied to continuous increasing ordinal functions beginning with the exponential function. By its nature, this proposed characterization of predicativity is impredicative, since it requires the impredicative concept of ordinal or well-ordering. It is supposed to be complete as looked at from the inside, in the sense

---

[2] On the face of it, this requires a second-order quantifier over all subsets X of the ordering up to a, but restriction to a first-order language can be maintained by taking X to be a predicate parameter.

that everything the "ideal predicativist" ought to accept – and nothing more – is eventually accepted, but it is certainly incomplete looked at from the outside, since the consistency of the limit system RA_γ with $\gamma = \Gamma_0$ is not provable in that system.

My subsequent work on predicativity branched along two paths, each carried out over an extended period. The first was to reformulate the characterization of predicative provability without any overt appeal to the impredicative notions of ordinal or well-ordering, so that it would describe more directly the expansion of reasoning that could be admitted by an ideal predicativist. The second was to see what part of mathematical practice may be accounted for in predicative terms.[3]

Along the first path, my rethinking of the formulation of systems for predicativity went through several stages, and was eventually conceived in the mid 90s as part of a much wider project, namely the determination of what I call *the unfolding of open-ended schematic systems*. The initial spark for that was provided by Kreisel's 1970 article, "Principles of proof and ordinals implicit in given concepts". He posed the general question: "What principles of proof do we recognize as valid once we have understood (or, as one sometimes says, 'accepted') certain given concepts?" As he elaborated it, "[t]he process of recognizing the validity of such principles (including the principles for defining new concepts, that is, formally, of extending a given language) is here conceived as a process of reflection ... . Granted that we have to do with an area [C] which lends itself to the kind of analysis indicated, it is evident that ordinals play a basic role. They index the stages in the reflection process." The two principal basic concepts considered by Kreisel were, in his terminology: 1. the concepts of ω-sequence and ω-iteration, and 2. the concepts of set of natural numbers and numerical quantification, the first being related to his earlier work (1960) on an autonomous progressions for finitist mathematics and the latter to mine (1964) on an autonomous progression for predicative mathematics. However, I decided instead that the formal systems considered for a given C should not be taken to involve the notions of ordinal or well-ordering in any way that is not already contained in the basic concepts of C. Moreover, I thought that extensions of set theory by certain axioms for

---

[3] See my survey article "Predicativity" (2005) for a much fuller description of my work on that subject, together with its background in the ideas and work of Poincaré, Russell, Weyl, Kleene and Kreisel.

"large cardinals" should serve as another possible example, in accordance with Gödel's view in his famous 1947 article on Cantor's continuum problem that the familiar systems such as ZFC "can be supplemented without arbitrariness by new axioms which are only the natural continuation of those set up so far."

The general notion of unfolding that I arrived at was first explained in my article, "Gödel's program for new axioms: why, where, how, and what?" (1996). It is applicable to formal systems in which schematic axioms and rules of inference are expressed using free predicate variables in an *expandable language* for which each expansion leads to new accepted substitution instances. The general questions raised for such open-ended schematic systems S are: *Which operations and predicates – and which principles concerning them – ought to be accepted if one has accepted S?* The answer for operations is straightforward: Any operation from and to individuals is accepted which is determined explicitly or implicitly (e.g., recursively) from the basic operations of S. Moreover, the principles which are added concerning such operations are just those which are derived from the way that they are introduced. The question concerning predicates in the unfolding of S is treated in operational terms as well: Which operations on and to predicates – and which principles concerning them – ought to be accepted if one has accepted S? For this, it is necessary to tell at the outset which *logical operations* on predicates are taken for granted in S. For example, in the case of *non-finitist (classical) arithmetic NFA*, these would be (say) the operations $\neg$, $\wedge$ and $\forall$, while in the case of *finitist arithmetic FA* we would be limited to positive propositional connectives and (in one formulation) the existential operator. Both of these have been investigated in collaboration with Thomas Strahm, to begin with in "The unfolding of non-finitist arithmetic" (2000), with the following results. We take the initial axioms for NFA to be the usual ones for 0, successor and predecessor (as the only constants and operations on individuals) together with the induction scheme

$$P(0) \wedge (\forall x)[P(x) \to P(sc(x))] \to (\forall x)P(x).$$

Further operations on individuals and predicates, and more elaborate axiom schemes are successively recognized via proofs of existence using the substitution rule. We showed that the operational unfolding of NFA is equivalent to PA, while the full (operational and predicate) unfolding is equivalent to predicative analysis, i.e. the union of the RA_α for $\alpha < \Gamma_0$.

In an unpublished MS in progress, "The unfolding of finitist arithmetic", Strahm and I have shown that both the operational and full unfolding of a system FA for finitist arithmetic are equivalent to the system PRA of Primitive Recursive Arithmetic. This supports Tait's argument in his paper "Finitism" (1981) that PRA represents the limit of finitist definitions and proofs, while it differs from Kreisel's claim (in his 1960 paper cited above, and elsewhere) that a system equivalent in strength to PA is its limit. I conjecture that a system of strength PA can be shown to be the unfolding of NF augmented by a suitable quantifier-free form of rules for definition and proof by induction on well-founded orderings. Finally, while there are no definitive results yet for the unfolding of set theory, a framework for that has been provided in my paper, "Operational set theory and small large cardinals" (2006a). There are other obvious candidates of open-ended schematic systems for which the unfolding notion ought to be investigated.

I can be somewhat briefer concerning the mathematical path in the work on predicativity. In Hermann Weyl's work *Das Kontinuum* (1918) of his predicativist period, he explained how all of 19th century analysis of piece-wise continuous functions could be accounted for in predicative terms. Examination of Weyl's system showed that it could be formalized within a theory of finite types conservative over PA. By modifying this to a more flexible system W of variable finite types also conservative over PA, I was able to verify that much of 20th century functional analysis of Lebesgue measurable functions can be formalized in W. I was then led to conjecture that *all of scientifically applicable mathematics can be formalized in W*, and hence rests on a completely predicative basis; see "Why a little bit goes a long way. Logical foundations of scientifically applicable mathematics" (1993). That conjecture has been verified to a considerable extent for the main fundamental results in functional analysis. This is relevant to the Quine-Putnam *indispensability thesis* that led them to accept substantial portions of impredicative set theory as seemingly inextricably necessary for science. As I have written op. cit., "By the fact of the proof-theoretical reduction of W to PA, the only ontology it commits one to is that which justifies acceptance of PA." Moreover, as is well-known, the latter is reducible to the intuitionistic system HA of Heyting Arithmetic, which does not require any platonistic ontology whatever. Thus, in my view, "if one accepts the indispensability arguments, practically nothing philosophically definitive can be said of the entities which are then supposed to have the same status – ontologically and epistemologically – as the entities of natural science." My conclusion was that the indispensability arguments are thus completely vitiated.

What is the proper role of philosophy of mathematics in relation to logic, foundations of mathematics, the traditional core areas of mathematics, and science?

How can one's choice of philosophy of mathematics dictate what it is right to do and say in mathematics, i.e. in its foundations? Consider the candidates on offer: Formalism, finitism, constructivism, predicativism, logicism, nominalism, fictionalism, instrumentalism, platonic realism, structuralism, modal structuralism, scientific naturalism, mathematical naturalism, and quasi-empiricism, among others, including some in competing subvarieties. For those thinkers who have arrived at what they take to be the one true philosophy, the answer goes without saying. Moreover, among the pioneers to our subject such as Cantor, Frege, Brouwer and Hilbert, that stance was very efficacious in leading to substantial research programs. However, as those programs were developed, along with great strides they were marked by serious difficulties. Comparable programs nowadays that are being vigorously pursued by a number of adherents are Martin-Löf's constructive type theory, the Bishop school of constructivity, the large cardinal program in set theory, and categorical foundations. For these, the difficulties are of a different nature. Like predicativity, the first two require radical restrictions of what is admitted to mathematics, while the large cardinal program makes use of a radical extension; finally, the categorical program claims to usurp logical foundations. Most logicians have not committed to such definite philosophical views, since active debate between the various positions makes a choice between them difficult, and radical solutions are discomfiting. Among mathematicians, there is a widespread view that ongoing current mathematics on the whole is more reliable than any of the philosophically motivated programs that have been proposed to replace it, and that the only foundations that need be considered (if any at all) is organizational.

My own view lies between these extremes. First of all, the historical development of mathematics shows that *not anything goes*, that a number of notions, assumptions and supposed results have been found to be seriously problematic at one stage or another, e.g. infinitesimals, imaginary numbers, points at infinity, trigonometric series expansions of arbitrary functions, probabilistic arguments, etc., etc. In the past, mathematicians dealt with these on a case by case basis. In my article, "Working foundations" (1993a), I have argued that outside of the grand foundational schemes, what logic has had to offer in these days is work that is "a direct continuation by more conscious, systematic means of foundational moves which have been carried on within mathematics itself from the very beginning." While not driven by any particular philosophical view, these *foundational ways* are often usefully informed by philosophical distinctions.

The second thing I have been concerned with at a philosophical level is a critical examination of several foundational schemes, including categorical foundations, Lakatosian quasi-empiricism, and the large cardinal program, in the articles (among others) "Categorical foundations and foundations of category theory" (1977), "The logic of mathematical discovery versus the logical structure of mathematics" (1981), and "Why the program for new axioms needs to be questioned" (2000), respectively. I have already mentioned another critique of this character above, in connection with the Quine-Putnam indispensability thesis. In all of these, while not espousing a fixed *positive* philosophical position, I have brought to bear some fairly strong *negative* views. Limitation of space prevents me from going into any detail here about the content of these critiques.

But let me enlarge on the positive vs. the negative aspects. Because of my substantial involvement over the years in studying the concept of predicative definability and provability, some have assumed that my philosophical position is that of predicativism; this is definitely not the case. As I have written in the preface to my collection of essays, *In the Light of Logic* (1998), rather, "I am a convinced antiplatonist in mathematics. ... according to the platonist philosophy, the objects of mathematics such as numbers, sets, functions and spaces are supposed to exist independently of human thoughts and constructions, and statements concerning these abstract entities are supposed to have a truth value independent of our ability to determine them. Though this accords with the mental practice of the working mathematician, I find the viewpoint philosophically preposterous ..." (or, as I have written elsewhere, set-theoretical platonism is the "medieval metaphysics of mathematics"). To go on, "[i]t should not be concluded from this, or from the fact that I have spent many years working on different aspects of predicativity, that I consider it the be-all and end-all in non-platonistic foundations. Rather, it should be looked upon as the philosophy of how we get off the ground and sustain flight mathematically without assuming more than the basic structure of the natural numbers to begin with. There are less clear-cut conceptions which can lead us higher into the mathematical stratosphere ... That such conceptions are less clear-cut than the natural number system is no reason not to use them, but one should look to see *where* it is necessary to use them and what we can say about what it is we know when we *do* use them."

What are the most important open problems in the philosophy of mathematics and what are the prospects for progress?

Here are some rather general questions:

(i) Is the Continuum Hypothesis a definite mathematical problem?

(ii) What is a natural system and what is the interpretability order of natural systems?

(iii) What makes mathematics such a distinctive body of thought? What determines what counts as mathematics and what doesn't?

(iv) Is the structure of mathematics essentially logical in nature? If not, what is it?

(v) Is the use of formal systems an adequate model of mathematical practice?

(vi) How is it that mathematics is so successfully applicable to natural science? Does that depend on what part of mathematics is being applied?

Concerning (i), I came to the conclusion some years ago that CH is an *inherently vague* problem (see, e.g., the article (2000) cited above). This was based partly on the results from the metatheory of set theory showing that CH is independent of all remotely plausible axioms extending ZFC, including all large cardinal axioms that have been proposed so far. In fact it is consistent with all such axioms (if they are consistent at all) that the cardinal number of the continuum can be "anything it ought to be", i.e. anything which is not excluded by König's theorem. The other basis for my view is philosophical: I believe there is no independent platonic reality that gives determinate meaning to the language of set theory in general, and to the supposed totality of arbitrary subsets of the natural numbers in particular, and hence not to its cardinal number. Incidentally, the mathematical community seems implicitly to have come to the same conclusion: It is not among the seven Millennium Prize Problems established in the year 2000 by the Clay Mathematics Institute, for which the awards are $1,000,000 each; and this despite the fact that it was the lead challenge in the famous list of unsolved mathematical problems proposed by Hilbert in the year 1900, and one of the few that still remains open.

I have been asked to explain what I mean by the statement of a problem being inherently vague. The idea is that, not only is it vague, but there is no reasonable way to sharpen the notion or notions which are essential to its formulation without violating what the notion is supposed to be about. For example, the notion of *feasibly computable number* is inherently vague in that sense. And, for the statement of CH, the notion of arbitrary subset of N can't be sharpened to arbitrary constructible subset of N, or any specific relativization thereof, without violating the idea of *arbitrary subset of a set*, independent of any means of definition. I think progress can be made on elaborating the idea of inherently vague notions; whether that can be used to strengthen the case that CH is an inherently vague problem remains to be seen.

Concerning (ii), much has been made by workers in the metamathematics of set theory of the observed fact that all natural systems extending ZFC that have been considered are comparable by the relation of relative interpretability in Tarski's sense. Moreover, in many cases systems based on quite different principles turn out to have the same interpretability strength. It happens that large cardinal axioms have in all such cases figured as an essential link in establishing the interpretation (see, for example, Steel 2000, p. 227). The central role of large cardinal axioms in these phenomena has been taken to suggest that that they in some way reflect a pre-established harmony. Be that as it may, the phenomenon of linear ordering under interpretability has been observed to hold much more generally for all natural systems that have been considered extending certain very weak subsystems of arithmetic (cf. Friedman 2007, Section 7), and large cardinal axioms don't have anything to do with that more general phenomenon. So, really, the underlying question here is: what is a natural system? And then, if there is a reasonable answer to that question, what is the interpretability order and related orders (translatability, consistency, etc.) between such systems? Friedman points out (loc. cit.) that it is not linear among algebraic systems, e.g. the theories of discrete linear order without end-points and dense linear order without end-points are incomparable. Nor is it linear among finite extensions of arithmetic (hence f.i.p. systems), as shown in Lindström 1997, Ch. 7. However, the latter systems are not natural because they are [3]cooked-up[2] by means of arithmetization techniques. It would be quite remarkable, and might be considered some sign of the inner harmony of mathematics, if all natural systems extending arithmetic turned out to be linearly ordered by interpretability. If one believes that this should be the case, a search for counterexamples among candidate explications could be a first step toward narrowing down the informal concept of natural system.

Problems (iii)–(v) are interrelated, and may be connected with problem (vi). Let me conclude with some ideas about these. Discovery in mathematics is one of the highest exercises of creative intelligence. But confirmation of mathematical discoveries requires rigorous calculation and demonstration, and in this respect mathematics is logical at its core. Moreover, mathematics is progressive, it builds on what came before. Thus, since there can be no infinite regress, from the point of view of logic mathematics must rest ultimately on some sort of axiomatic foundations. While mathematicians may accept this in principle, there is a sharp dichotomy between the logicians' conception of mathematics and that of the practicing mathematician. The latter pays little or no attention to logical or foundational axioms, even if he or she subscribes to some overall foundational viewpoint such as that of axiomatic set theory. And in fact, the logical picture of mathematics bears little relation to the logical structure of mathematics as it works out in practice. The use of certain basic structures like the natural numbers and the real numbers (and of structures built directly from them like the integers, rationals and complex numbers) is ubiquitous, and there is constant appeal to such principles as proof by induction and definition by recursion on the naturals and of the lub principle for the real numbers. But these are not viewed from an axiomatic point of view, e.g. from that of the Peano Axioms for the naturals. The essential difference is that the language of PA

is limited to a fixed vocabulary, whereas induction and recursion can be applied in *any* subject in which natural numbers play some sort of role. For example, the operation x^n is defined in any (multiplicative) semi-group for every element x and natural number n, and its properties are proved by induction on n. So even where the practicing mathematician invokes the basic axioms of the natural numbers, that is done without restriction to a fixed vocabulary. According to the current set-theoretical point of view, all such concepts that the mathematician might want to use in addition to those expressed in PA are defined in the language of ZFC, so we need only look no further in order to give full logical scope to what underlies daily mathematics. It seems however, that if we accept the language of set theory we ought to accept notions *not* defined in that language, such as the notion of truth in the set-theoretical universe. Moreover there are informal outlying notions that have mathematical coherence, but are not (as given) defined within set theory. One such, for example, is the notion of free choice sequence used by the Brouwerian intuitionists; this is separate from the fact that a formal system for f.c.s.'s can be modeled in set-theoretical terms. Another example of a mathematical notion that is not set-theoretically defined is the informal concept of randomness applied in various contexts, though again axiomatics of randomness has been modeled set-theoretically. Finally, when mathematics is applied to natural science, it makes direct use of physical concepts like force, mass, charge, etc., etc., that are evidently not expressed in set-theoretical terms at all. In my 2005 ASL lecture, "Open-ended schematic axiom systems" (2006), I have proposed an informal framework to account for mathematical practice and its actual and future possible applications in a more direct way than through the use of the various formal systems currently dominating logical work. This is work in progress, as an extension of my earlier work on unfolding of open-ended schematic systems. An essential new feature is the introduction of a quite general underlying "proto-mathematical" framework for operations and properties; that allows for the interaction of basic schematic systems like those for the natural numbers, real numbers, and subsets of any domain. I believe the emphasis on conceptual open-endedness will also provide a new perspective on the phenomenon of incompleteness which was the preoccupation above.

References

Anita Burdman Feferman and Solomon Feferman

— (2004), *Alfred Tarski: Life and Logic*, Cambridge University Press, New York, NY.

Solomon Feferman

— (1960) Arithmetization of metamathematics in a general setting, *Fundamenta Mathematicae* 49, 35–92.

— (1962) Transfinite recursive progressions of axiomatic theories, *J. Symbolic Logic*, 259–316.

— (1964) Systems of predicative analysis, *J. Symbolic Logic* 29, 1–30.

— (1977) Categorical foundations and foundations of category theory, in *Logic, Foundations of Mathematics and Computability Theory* (R.E. Butts and J. Hintikka, eds.), vol. 1, Reidel, Dordrecht, 149–169.

— (1981) The logic of mathematical discovery versus the logical structure of mathematics, in *PSA 1978*, Philosophy of Science Assoc., East Lansing, 309-327; reprinted as Ch. 3 in (1998), 77-93.

— (1989) Finitary inductively presented logics, in *Logic Colloquium '88* (R. Ferro, et al., eds.), North-Holland, Amsterdam, 191-220.

— (1993) Why a little bit goes a long way. Logical foundations of scientifically applicable mathematics, in *PSA 1992*, Vol. II, 442-455; reprinted as Ch. 14 in (1998), 284-298.

— (1993a) Working foundations '91, in *Bridging the Gap: Philosophy, Mathematics and Physics*, (G. Corsi, et al., eds.), Kluwer, Dordrecht, 99-124; reprinted as Ch. 5 in (1998), 105–124.

— (1996) Gödel's program for new axioms: why, where, how and what?, in *Gödel '96* (P. Hajek, ed.), *Lecture Notes in Logic* 6, 3-22.

—— (1996a) Three conceptual problems that bug me, http://math.stanford.edu/~feferman/papers/conceptualprobs.pdf (Unpublished lecture text for 7th Scandinavian Logic Symposium, Uppsala, 1996).

—— (1997) My route to arithmetization, *Theoria* 63, 168–181.

—— (1998) *In the Light of Logic*, Oxford University Press, New York, NY.

—— (2000) Why the program for new axioms needs to be questioned, in the symposium "Does mathematics need new axioms?, (S. Feferman, H. M. Friedman, P. Maddy and J. R. Steel), *Bull. Symbolic Logic* 6, 401-413.

—— (2005) Predicativity, In *The Oxford Handbook of Philosophy of Mathematics and Logic* (S. Shapiro, ed.), Oxford University Press, Oxford, 590–624.

—— (2006) Open-ended schematic axiom systems, (abstract), *Bull. Symbolic Logic* 12, 145. (Unpublished lecture for the Annual Meeting of the Assoc. for Symbolic Logic, Stanford, March 19–22, 2005).

—— (2006a) Operational set theory and small large cardinals, http://math.stanford.edu/~feferman/papers/ostcards.pdf, (to appear in the Proceedings of the WoLLIC conference, Stanford, July 2006).

Solomon Feferman and Clifford Spector

—— (1962) Incompleteness along paths in progressions of theories, *J. Symbolic Logic* 27, 383–390.

Solomon Feferman and Thomas Strahm

—— (2000) The unfolding of non-finitist arithmetic, *Annals of Pure and Applied Logic* 104, 75–96.

—— (2001) The unfolding of finitist arithmetic (abstract), *Bull. Symbolic Logic* 7, 111–112.

Harvey Friedman

___ (2007) Interpretations, according to Tarski, http://www.math.ohio-state.edu /~friedman/ pdf/Tarski1,052407.pdf.

Georg Kreisel

___ (1960) Ordinal logics and the characterization of informal concepts of proof, *Proc. of the International Congress of Mathematicians, 14-21 August 1958*, 289–299.

___ (1970) Principles of proof and ordinals implicit in given concepts, in *Intuitionism and Proof Theory* (A. Kino et al., eds.), North-Holland, Amsterdam.

Per Lindström

___ (1997) *Aspects of Incompleteness*, Lecture Notes in Logic 10.

John R. Steel

___ (2000) Mathematics needs new axioms, in the symposium "Does mathematics need new axioms?", (S. Feferman, H. M. Friedman, P. Maddy, and J. R. Steel), *Bull. Symbolic Logic* 6, 422–433.

William Tait

___ (1981) Finitism, *J. of Philosophy* 78, 524–546.

Alan Turing

___ (1939) Systems of logic based on ordinals, *Proc. London Mathematical Society* (2) (1939) 161-228; reprinted in Turing (2001).

___ (2001) *Mathematical Logic* (R. O. Gandy and C. E. M. Yates, eds.), *Collected Works of A. M. Turing*, Elsevier, Amsterdam.

Hermann Weyl

___ (1918) *Das Kontinuum. Kritischen Untersuchungen über die Grundlagen der Analysis*, Veit, Leipzig.

___ (1987) *The Continuum. A Critical Examination of the Foundations of Analysis* (English translation of Weyl 1918, by S. Pollard and T. Bole), Thomas Jefferson Press, distributed by Univ. Press of America, Latham.

11

Bob Hale

Professor of Philosophy
University of Sheffield, UK

Why were you initially drawn to the foundations of mathematics and/or the philosophy of mathematics?

Although, while an undergraduate in philosophy at Bristol, I attended various lectures on the foundations of mathematics – including one beautifully clear talk by Alfred Tarski – and read Stephan Körner's introductory book on the philosophy of mathematics, it was not until well into my postgraduate studies in Oxford that I started to get seriously interested in the subject.

Two occurrences were primarily responsible for this. First, an argument with a fellow student there led me to read Michael Dummett's paper 'Truth', and although it was quite some time before the full significance of its central ideas sunk in, I think it made me realize that issues in the philosophy of mathematics were not just of narrowly specialist interest, but could be important for ostensibly more central areas of philosophy—such as epistemology, metaphysics, and philosophy of language. Second, and encouraged by reading his paper on truth, I attended Dummett's lectures on Frege's philosophy of mathematics, which must have been an early version of several of the chapters in the book he published much later. Their effect was immediate, profound, and electrifying. Although I continued to work, with my supervisors – Gilbert Ryle and, later, David Pears – on topics in the philosophy of mind, I soon realized that the questions in philosophy which most interested me—about what kinds of things there are, and how we can know about them—came up in a very sharp and pressing form in connection with mathematics.

Even so, it was not until after I had been teaching, in Lancaster, for nearly a decade, that I started to work seriously on questions in the philosophy of mathematics. Around this time, I was fortunate enough to have regular opportunities to talk shop with Crispin Wright, whom I'd met in Oxford but not seen for a few years. He was then working on *Frege's Conception of Numbers as Objects*, and I on some related questions about Frege and singular terms. Crispin quickly persuaded me that I should start work on what became *Abstract Objects*. I can't recall ever having been more excited—or exhausted—by doing philosophy with someone as I was then. Most readers will know what happened next!

What examples from your work (or the work of others) illustrate the use of mathematics for philosophy?

In an obvious sense, mathematics sets the agenda for philosophy of mathematics, by providing – especially in fundamental mathematical theories such as arithmetic, analysis, and set theory – the data for which we are trying to give a philosophical account or explanation. At the most basic level, we want to give philosophically satisfying and defensible answers to such simple-seeming – but, as we all know, horribly difficult – questions as: *What is mathematics about? How can we know about it?*

Then, of course, there are more specific ways in which mathematical work can be especially important in philosophy. Most obviously, there are specific mathematical results—the Gödel incompleteness theorems and the Gödel-Cohen independence results being perhaps the most obvious examples—whose philosophical significance needs to be assessed, perhaps because they may have a direct bearing on the viability of particular philosophical accounts of mathematics, perhaps because they reveal gaps in our understanding of important mathematical concepts, or perhaps for yet other reasons.

There is also a much more routine way in which mathematical work may be useful – and, indeed, actually necessary – in philosophy. A would-be philosophical foundation for a given part of mathematics may be no good for straightforwardly technical reasons—in the most dramatic case, it may be inconsistent. But even if not actually inconsistent, it may be no good because it does not provide for the proof of the requisite theorems. Verifying these things is itself, at least in part, a mathematical task. A very modest example in my own work is the neo-Fregean construction of the real numbers I put forward a few years ago—but I am sure there are much better examples in the works of others.

What is the proper role of philosophy of mathematics in relation to logic, foundations of mathematics, the traditional core areas of mathematics, and science?

This is perhaps the most controversial of your questions. As I said in answering the previous question, mathematics provides the data for philosophy of mathematics, about which we can ask such questions as: *What is it about? How can we know about it?* But what we take a given part of mathematics to be about depends upon whether we take statements belonging to it as strictly and literally true—true when they are taken at face-value, in accordance with their surface grammar (or simply true, as I'll say for brevity). If we take the statements of elementary arithmetic at face-value, they involve simple and complex terms which purport to stand for objects, and quantifiers binding variables ranging over such objects. So if the accepted statements of arithmetic are simply true, arithmetic is about a certain collection of objects—the natural numbers. Even at this very elementary level, philosophers disagree quite sharply, with many of them rejecting a face-value construal and advocating alternative views which either try to hang onto the idea that arithmetical statements are true, while denying that they are really about what they appear to be about, or retain a face-value ('platonistic') interpretation, but deny that such statements are ever (non-vacuously) true. These alternative views, which include various forms of nominalism, fictionalism, and also some kinds of structuralism, are not intrinsically, or usually, revisionary as regards *mathematics*—they accept standard arithmetic—rather, they offer revised accounts of what accepting it involves, or should involve.

This suggests a further and more fundamental question underlying the questions so far highlighted, viz. *How much of mathematics can be accepted as simply true?* (I assume that 'can' implies 'ought' in this case!). And, unless the answer to this question is: *All of it*, we shall also need to ask: *What account is to be given of those parts which can't be accepted as simply true?* I think there is some sort of presumption in favour of taking any given part of mathematics as simply true, unless there are compelling reasons to do otherwise. Of course, it is hardly uncontroversial what should count as a compelling reason to do otherwise, and whether there are any such reasons. Crude nominalism takes the fact that, taken at face-value, certain mathematical theories imply the existence of abstract entities—usually infinitely many such, but even a few is too many from this point of view—to be *in itself* a compelling reason for refusing to take those theories as simply true. More plausible forms of nominalism take commitment to abstract entities to be problematic because they think it is difficult or even impossible to explain how we can make reference to such entities, or have any knowledge about them. Although I don't share this view, I do accept a general methodological principle underlying it—that if taking a given part of mathematics to be simply true really does preclude any credible account of how we can make reference to its ostensible objects, or know or form justifiable beliefs about what it says about them, that is a strong reason for rejecting a face-value interpretation. This is, of course, implicit in the central dilemma of Paul Benacerraf's highly influential paper 'Mathematical Truth' (*J.Phil.* 1973).

As a matter of fact, I believe that we *can* take elementary arithmetic at face-value, and accept it as true, and that the same goes for analysis. Beyond that, I do not have a settled and worked-out view, let alone anything I would consider a good argument for it. In particular, my view about set theory is not settled. I do not think the iterative conception of sets provides a philosophically satisfactory rational or justification for the axioms of standard set theory (*ZFC*)—in part because it seems to me to rest on an indecently obscure idea about the (metaphysical) dependence of sets on their members, but perhaps more importantly, because it merely postpones the difficult epistemological questions to which the axioms of set theory give rise. I would like to believe that an alternative basis for a quite powerful set theory can be given, along neo-Fregean abstractionist lines. But even if we can get a theory close to standard set theory which we can accept as simply true in this or some other way (and a fortiori, if we can't), it seems to me very likely that there will remain a substantial portion of set theory (involving stronger axioms which, for example, would settle the Continuum Hypothesis and Generalized Continuum Hypothesis one way or the other) which can't be accepted as simply true. So we shall face a difficult question about what attitude to take towards it.

To sum up what I've said thus far: a central question is—How much of mathematics can be accepted as simply true? It is quite possible that the most defensible answer is: Some, but by no means all of it. Accepting that answer would not, as such, involve adopting a revisionary attitude towards mathematics, as opposed to claims about the alethic and epistemic status of certain parts of mathematics. It might involve denying that certain mathematical theories aren't simply true, and perhaps that they can't be known—but it wouldn't necessarily mean rejecting or revising those mathematical theories themselves.

Could there be compelling philosophical reasons to revise mathematics itself? Could there be such reasons for revising logic in ways that would entail revision of mathematics? I think many philosophers effectively treat existing mathematics – or more accurately, the mathematics accepted by the vast majority of mathematicians – as sacrosanct, and so do not take very seriously positions, such as that of the intuitionists and their sympathizers, which call its fundamental principles into question, with revisionary consequences. I do not myself think one can dismiss the possibility of such revision. In particular, I believe a strong case for revision of classical logic has been made, mainly in the writings of Michael Dummett, but also in the work of other philosopher-logicians of a constructive tendency, such as Dag Prawitz and Per Martin-Löf. It may be that this case is not ultimately compelling, but at least it requires to be answered, and I do not myself believe I know how to answer it. I don't think one can uphold classical logic just on grounds of simplicity, as some of its defenders seem to suppose.

What do you consider the most neglected topics and/or contributions in late 20th century philosophy of mathematics?

Several questions which seem to me not to have been given as much attention as they deserve have to do with the *applicability* of mathematics. Of course, everyone agrees that any adequate philosophy of mathematics needs to account for the possibility of its application. It is true, also, that a great deal of impressive technical work has been carried out, in measurement theory for example, which is precisely aimed at explaining the application of a central part of mathematics – the system of real numbers – to empirical domains which ostensibly lack crucial structural properties of the system of real numbers itself. But the kind of explanation there provided, which involves establishing a homomorphism between the mathematical and empirical domains, cannot get off the ground without crucial steps of idealization in the representation of the empirical domain. At the level of language, this process of idealization involves a transition between the essentially vague or inexact terms and concepts we use to describe the world of experience and experiment, on the one side and, on the other, the exact terms and sharp concepts of mathematics. While it would be quite wrong to suggest that the problems concerning the notion of idealization have been neglected by philosophers of science and of mathematics, it seems to me that the problems are still with us. Quite fundamental philosophical questions in this area – including: What is idealization? and: How is it justified? – have yet to be given satisfactory answers.

I think these problems arise on any approach to the philosophy of mathematics, but it is perhaps worth adding that they arise in a very direct way on the abstractionist approach to which I am myself attracted. On this approach, one tries to use abstraction principles—principles of the form:

$$\forall\alpha\forall\beta(\Sigma(\alpha) = \Sigma(\beta) \leftrightarrow \alpha \approx \beta)$$

where $\approx$ is an equivalence relation on entities of the type of α,β, ..., and Σ is a function from entities of that type to objects—as a means of implicitly defining the fundamental concepts of a mathematical theory in such a way that its basic laws may then be derived, given only a suitable background logic. The abstraction principles invoked may be second- or higher-order, involving, say, second-level equivalence relations defined over first-level properties, as with Hume's principle:

$$\forall F\forall G(\text{the number of } Fs = \text{the number of } Gs \leftrightarrow F \text{ and } G \text{ are 1-1 correlated})$$

on which abstractionists wish to build arithmetic. But to see the potential difficulty whose solution seems to require idealization, it is better to focus on first-order abstraction, in which the equivalence relation is a relation on objects. Suppose one wished to introduce lengths – conceived as a species of *abstract* objects – by means of abstraction. One might naturally propose:

$$\forall x\forall y(\text{the length of } x = \text{the length of } y \leftrightarrow x \text{ is as long as } y)$$

where the bound individual variables are taken to range over concrete objects, such as planks, steel rods, etc. But then we must ask what relation is denoted by 'as long as'. It seems that any relation for which we can determine empirically whether it holds between given concrete objects must fail to be transitive, and so must fail to be an equivalence. This suggests that abstractionist techniques can get no grip at this very basic level—at which we presuppose no abstract objects, but are dealing with relations among just concrete objects—without appealing to idealization or something very like it. This does not necessarily raise any difficulty for the use of higher-order abstraction. But I think it does probably bear quite closely on the prospects for developing an abstractionist treatment of fundamental geometrical concepts. However, I cannot go further into these issues here. Instead, I want to mention one other question concerning applicability which seems to me to deserve more attention.

As is well known, Frege held that it is only applicability which sets arithmetic apart from mere games and raises it to the level of a science. But Frege went further, by claiming – to put it somewhat loosely, and not in his own words – that acceptable definitions of fundamental mathematical concepts, such as cardinal or real number, should somehow provide for their possible applications. Of course, this idea needs more careful formulation before it can usefully be discussed—obviously, Frege did not mean that each specific application of elementary arithmetic or analysis should somehow be anticipated in satisfactory definitions of *natural number* or *real number*. As Dummett, I think, puts it somewhere, what Frege is more charitably supposed to have thought is rather that from good definitions of these concepts, it should be possible to extrapolate the general principle(s) governing, or providing for, their applications. Whether this claim – which Crispin Wright has labelled 'Frege's Constraint' should be accepted – even in those cases in which Frege certainly applied it, is controversial. Wright himself makes a case for accepting the constraint as applied to *natural number*, whilst doubting that it can be justified in relation to *real number*. I am inclined to endorse the constraint in both of these cases; but I would immediately agree that there are many mathematical concepts for which it is quite obviously inappropriate. The underlying question which I think merits further attention is: what makes the difference between those concepts for which Frege's Constraint seems reasonable – those where it is plausible to think that applicability is somehow of the essence – and those where it is not?

I don't of course claim that there aren't many other topics which deserve more attention than they've received in recent philosophy of mathematics. These are just some which readily occur to me.

What are the most important open problems in the philosophy of mathematics and what are the prospects for progress?

I really don't have much to add here to what I've said already. My questions about the subject-matter of mathematics and its epistemology – and my further question about how much of mathematics can be accepted as simply true – are still open, at least in the sense that there are no generally agreed answers to them. I wouldn't go so far as to claim that they are the most important problems or questions, but they are surely the first, and most basic, questions we ask. But as experience shows, they are certainly not the first questions for which we may expect to get agreed answers. Indeed, experience – I mean the experience of working in the philosophy of mathematics – may lead one to doubt that we shall ever reach agreement about them. That is not to say that we don't – much less that we can't – make progress. In practice, most of us start out with some allegiance to a particular approach, or kind of answer, to one or other of the basic questions, and work it through—until we run into some insuperable obstacle or at some criticism we can't see how to meet. Along the way, we learn things about what won't work, or about the limitations of this or that approach.

12
Geoffrey Hellman

Professor of Philosophy
University of Minnesota, USA

Early Interest in Foundations of Mathematics and Philosophy of Mathematics

My father was an admirer of Bertrand Russell and spoke of his work with Whitehead, *Principia Mathematica*, as well as several other books by Russell in philosophy and human affairs, in glowing terms as I was growing up. At some point while I was in high school, I recall his giving me a book of Russell's essays which included "What is Number?", which I remember reading with great interest. I was amazed to learn that we finally *knew the answer* (*sic*) to this great question, that a number is just a class of (all) equinumerous classes, as Russell said he had learned from an obscure (to me, then) German logician named Frege. I didn't really question the answer, but I remember wondering how we knew that it was right. A bit later, still in high school (a rural public one in New Jersey) and trying to teach myself basic calculus from a textbook I had picked up, I came upon a section about functions. It said that in general a function is just a set of ordered pairs where the second member of each pair is uniquely determined by the first. I remember saying to myself, "Ok, but why? Isn't a function some sort of rule indicating a dependence of one thing on another? Don't you need a formula?" I don't think such questions kept me up at night back then; probably I filed them away somewhere, focusing on the various operations and problems I had to learn in the meantime. But I surmise from later developments that they must have subtly gnawed away toward consciousness in my sub- or semi-conscious mind. Later, in college (at Harvard), I learned about Gödel's incompleteness theorems from Quine's discussion at the end of *Methods of Logic*. After studying those theorems in detail (in a course of Burton Dreben's), I was hooked. Further courses from Hilary Putnam and then Michael Dummett at

Oxford (during a fellowship year abroad while I was in graduate school at Harvard) guaranteed that I "stayed hooked".[1]

Examples illustrating the use(s) of mathematics for philosophy

Without any claim to exhaustiveness, I would call attention to four principal ways in which mathematics is useful for philosophy:

1. Focusing attention on core questions of metaphysics and epistemology, forming the backbone of philosophy of mathematics itself;

2. Highlighting problems concerning the applications of mathematics in the sciences;

3. Exemplifying pluralism in mathematics, both in methods and substance;

4. Furnishing paradigms of successful explication, analysis of concepts, and theory construction.

Under 1. fall the big questions of philosophy of mathematics: What is it about, if anything? How should mathematical existence be understood? What sort of knowledge does it provide, if any? What portions are analytic (if any), *a priori*, necessary? Is any synthetic *a priori*? How do we manage to grasp the conceptual apparatus of mathematics, especially when it seems to transcend all possible experience? In particular, how do we understand various concepts and levels of infinity? What does this tell us about meaning and communication? And so forth.

Especially important in connection with these questions are the following areas and topics: geometry, non-Euclidean as well as Euclidean, and the distinction between pure and physical geometry; the number systems and how they are established; abstract algebra, including category theory: what if any substantive axioms—as assertions, not merely structural defining conditions—are at work? Extraordinary mathematics, including higher set theory and some of topos theory; how can strong axioms of infinity (from Replacement through the small- and large-large cardinal axioms)

---

[1] Andrew Carnegie, commenting on President Theodore Roosevelt's "trust busting", complained: "We bought that president; the trouble is, he didn't stay bought!"

be justified? To what extent are such new axioms implicated in mathematical problems closer to ordinary practice (as suggested by work of Harvey Friedman)?

Under 2. arise challenging questions such as whether the applicability of mathematics to the material world (especially functional analysis through physics) is "genuinely mysterious", really eluding "naturalistic explanation", etc. (as claimed, for instance by Mark Steiner), or whether this is something of an illusion arising from ignoring many failures or from other sources (e.g. counting as irrational or non-rational relying on "merely" successful past practices, i.e. failing to take to heart one of Goodman's lessons in *Fact, Fiction, and Forecast*). Along different lines, given Weyl's dictum that mathematics must serve the needs of the natural sciences, arise questions concerning the strength and limitations of systems of mathematics weaker than standard, classical ones, such as intuitionistic mathematics, Bishop constructivism, recursive analysis, predicative analysis, and so on. Despite various advances made on behalf of these systems, limitations appear to arise regarding the first two (based on intuitionistic logic) in connection with extremization problems (extreme value theorem and cognates, existence theorems relating to Plateau's problem, etc.). (See response to Qu. 5, below.) Other challenges concern domains of operators (e.g. a theorem of Pour-el and Richards that characterizes the class of computable linear operators on a Banach space as the bounded ones; unbounded operators—e.g. those for position or linear momentum in ordinary quantum mechanics—fail to be constructively defined at vectors within the classical domains (already less than the whole space, by a classical theorem (Riesz and Nagy, §114). Limitations on all these relatively weak systems (compared to fully classical ones) may also arise regarding uses of *non-separable spaces* (lacking a countable basis) in theoretical physics (in connection with systems involving infinitely many degrees of freedom). Here what is challenged is the attempt to get by with *countable* mathematics. Although genuine reliance on the uncountable is no doubt exceptional, claims of adequacy of a given system for the sciences don't require many counterexamples to refute. Furthermore, as a lesson from philosophy of science, a broad conception of "applications to science" to include high level theory as well as experiment is in order. (See response to qu. 5, below.)

Regarding pluralism, the weaker, non-classical systems just alluded to, corresponding to various constructivist approaches to mathematics, serve as examples, at least on a "liberal", as opposed to a "radical", conception of constructivism, which allows for "peaceful coexistence". On this conception, classical non-constructive statements and reasoning are not regarded as somehow deficient in meaning, but simply as less informative in certain respects than constructivist reasoning. We may not care about this in certain contexts, e.g. in investigating models of spacetime structure, where we want to know what *would* be the case given that certain prior conditions held regardless of whether an ideal computing agent could come to demonstrate it. In other contexts, retaining "computability" or "constructive computability" will be desired or even required, and then a constructivist system will serve us well. The appearance of irreconcilable conflict between classical and, say, intuitionistic analysis is illusory: The logical symbols differ systematically in meaning across the two systems (as both sides can readily see), and so really should be subscripted to avoid confusion. (Dummett writes (in his *Elements of Intuitionism* 1977) as if intuitionist theorems that look like negations of classical ones really are, but if I was right in my "Never Say 'Never'!" (*Phil. Topics* 1989), this is simply mistaken.) Genuine disagreement between classicism and *radical* constructivism is not at the level of mathematics itself but at the level of metamathematics or philosophy. Since certain classical truth conditions, regarding e.g. non-constructive existence statements, cannot be put in verificationist terms, they are devoid of communicable meaning, according to the radical constructivist. The classicist rather applies *modus tollens*, seeing in such cases counterexamples to a verifiability theory of meaning.

An especially interesting example of pluralism is presented by *smooth infinitesimal analysis* (part of *synthetic differential geometry*), which bases calculus on nil-square (and nil-potent) infinitesimals, avoiding contradictions by restricting reasoning to intuitionistic logic. ("Not all quantities are either equal to 0 or distinct from 0" is actually demonstrable in SIA which has models provided in topos theory.) Here forswearing use of the law of excluded middle cannot be traced to constructive truth (i.e. proof) conditions, but rather arises along with restriction of function spaces to "smooth worlds", successfully modeled in certain toposes where an internal logic generalizes classical bivalent logic resulting in intuitionistic logic. I tried to analyze this situation as a case of (at least the *possibility* of) genuinely vague objects (the nil-square infinitesimals), subject nevertheless to mathematically rigorous reasoning.

(See my piece in the *Journal of Philosophical Logic*, 2006.) The case is more challenging for the philosopher of mathematics than the better known one of non-standard analysis, since the latter retains classical logic and never derives anything that even appears to contradict classical analysis. (The interested reader should consult John Bell's excellent *Primer of Infinitesimal Analysis*, which develops SIA with applications to physics in some detail.)

To what extent and in what respects do set theory and topos theory present conflicting views of mathematics? Is peaceful coexistence possible? How is pluralism in mathematics similar to and/or different from pluralism in the natural sciences? These and related questions require our attention and reflection.

Turning to the fourth item of our list, mathematics since ancient times has been regarded as paradigmatic for philosophy. It still is, despite all the intervening developments in both fields. Nowadays, however, we are more circumspect in how we describe the paradigms. Rather than simply taking mathematics generally as exemplifying "certain knowledge", we would restrict this, perhaps severely to, say, primitive recursive arithmetic or some other weak system, e.g. an elementary theory of finite sets, perhaps. In the wake of Gödel's incompleteness theorems, we recognize limits to "certainty" regarding, e.g., consistency claims of powerful theories and (in light of reverse mathematics) regarding many key theorems, e.g. the Bolzano-Weierstrass theorem (equivalent over a weak base theory to the relatively non-controversial arithmetical comprehension axiom). And when it comes to strong axioms of set theory, such as Replacement, or stronger, we don't demand anything like "self-evidence", but tend to invoke informal notions such as *coherence* of the idea of a set-theoretic universe that extensive, or analogies with the systematizing role, sometimes even "explanatory role", of theoretical postulates in the foundations of physics, such as fundamental symmetries or conservation principles.

A more modern way in which we regard mathematics as paradigmatic concerns the construction and elaboration of some of its important concepts. Take, for example, "continuity" or "smoothness". These are widely used presystematic or informal notions made precise in various ways within mathematics. The precise notions capture many clear cases according to informal usage, both positive and negative, but many new cases (including pathologies) arise that were never previously contemplated. Exactly for this reason, what informally is regarded as an obvious platitude, say, that a continuous function on the unit interval which is negative at 0 and positive at 1 has a zero (the Bolzano case of the intermediate value theorem), while convincingly illustrated on graph paper, actually requires a rigorous analytic proof (as Bolzano himself realized). Indeed, the standard argument is not even constructive: a function's having a zero can be a long way from *our* having any effective method of "finding" a zero. (Indeed. constructivization requires modifying the statement in one way or another.) And along with this, the presystematic notion splits into a variety of notions, e.g. piecewise continuity, uniform continuity, etc., along with a hierarchy of differentiability requirements all the way up to smoothness (C^∞). Moreover, we learn that metrical notions (inherent in the epsilon-delta definitions) are not essential, that continuity can be treated as a topological concept. Thus whole bodies of theory are developed interrelating whole batteries of concepts in new and insightful ways. Moreover, as if this weren't enough, modern mathematics has generated an *embarras de richesse*: as broached in the discussion of pluralism above, mathematicians have brought category to bear on a radically different approach to continuity and smoothness, reviving and making rigorous earlier ideas of infinitesimals, providing a "non-punctiform analysis" as an alternative to the standard, classical method of limits. In sum, the philosopher is right to be impressed—and even just a bit envious.

Role of Philosophy of Mathematics (POM) in relation to Logic, Foundations of Mathematics (FOM), etc.

There is little point in attempting to draw boundaries between these fields. They are in any case not sharp and no doubt change over time due to various factors, some quite adventitious. As the anthropologist, Marvin Harris, once said, "There is nothing in nature quite so separate as two mounds of expertise."

As currently practiced, POM is at the same time continuous with logic and FOM and broader than them in its efforts to reflect on (among many other things) the significance of their formal results for understanding epistemological and metaphysical issues arising from mathematics, both pure and applied. Thus, for example, precise delineation of degrees of computability belongs to logic; investigations of the consistency strength of various formal systems useful for codifying various portions of mathematics belong to both logic and FOM; while elucidation of implications of limitative results such as the Gödel incompleteness theorems regarding broader issues such as "mathematical truth", "knowability of consistency", or "mechanism", would be regarded as belonging to POM. Note, however, that the epistemological significance of certain bodies of work in logic or FOM is so obvious—e.g. results of reverse mathematics or of predicative analysis or of intuitionistic type theory, etc.—that they could with justification be said to belong to POM as well, pretty much as they stand, with research papers on these topics amplified slightly, if need be, by an introductory and concluding paragraph, respectively providing philosophical motivation and drawing some (perhaps nuanced, qualified) philosophical conclusions. What I'm driving at here is that such formal results are part and parcel of the task of mapping out relations of epistemic grounding of many important mathematical results, showing in a precise way "what rests on what" (a nice title of one of Feferman's papers), or what is indispensable for what. The fact that such work might be accomplished by logicians with appointments in mathematics departments rather than philosophy departments would hardly mean that the work itself was not philosophical in its actual content. (As it happens, both Friedman and Feferman hold or held joint appointments in mathematics *and* philosophy departments—to the credit of both departments at both institutions—but my point will be clear enough in any case.)

That said, there is undoubtedly a greater endurance among philosophers for continual probing of metaphysical and epistemological issues than among logicians and mathematicians whose bread and butter (including promotions I suppose) literally depend on theorems. The philosopher, of course, may prove the occasional theorem as well, but that will normally be part of a philosophical investigation, such as a reconstruction program aimed at showing the adequacy of some particular system for encompassing a certain range of mathematical results, e.g. a system susceptible of a nominalistic interpretation or one designed to express a structuralist standpoint, and so forth. Typically, such systems are proposed as ways of avoiding some apparent commitments of ordinary practice seen as problematic (e.g. to *abstracta* or axioms (e.g. Infinity, Replacement, or Choice in set theory) seen as hard to justify, or to implicit uniqueness claims that seem arbitrary or untrue, etc.). They are proposed, that is, neither as literal substitutes for better entrenched systems or methods of actual mathematical practice ("revolutionary", in the terminology of Burgess and Rosen, or what I dubbed "Maoist mathematics" in my review of their book, *A Subject with No Object*, in *Philosophia Mathematica* (1998) 357-68), nor as revealing what working mathematicians have really meant all along ("hermeneutic", as Burgess and Rosen call it), but rather as rational reconstructions that aim to show certain ways in which mathematical results *can* be justified, or more readily justified by avoiding certain philosophical or linguistic or logical problems. "Either revolutionary or hermeneutic" is simply a false dilemma. If anything, to the extent that it is successful, a reconstruction program (say, along nominalist or structuralist lines), goes some distance toward sustaining mathematical practice as it stands, without thereby implying that it exposes "the true meaning" of ordinary mathematical language. It shows that such language need not be the final arbiter of what is presupposed in mathematical or scientific work, or better, what really needs to be presupposed. The mathematician (or logician, or philosopher, or layperson), that is, is being invited to think on at least two different planes: one that of ordinary practice, another according to a reconstruction that captures the essentials of the mathematics (or, in cases such as Field's instrumentalism, the essentials of the mathematics' applications to empirical science), without certain problematic commitments. The latter are then demoted to the level of *façon de parler*, part of a useful way of speaking but not *the* governing source of logical regimentation. This runs counter to Quinean methods, taken over implicitly by Burgess and Rosen, whereby direct paraphrases of ordinary (mathematical) language in the privileged formal language of first-order logic with equality serve to indicate e.g. ontological commitments via the '$\exists$' of that formalism. I would submit that such "literal paraphrases"— or "face-value readings"—are often not worthy of the name; they are not necessarily reliable guides as to "genuine commitments", i.e. they should not be read as fulfilling the "hermeneutic" approach. The point is that working mathematicians typically have not reflected on exactly *what* their commitments are, or, if they have, they have come to many different individual conclusions, some very far from a "face-value reading". For a striking example, Tarski surely spoke the informal language of mathematics (allowing Polish as well as English) as well as anyone but was a convinced nominalist (as the Fefermans report in their biography). Regarding axioms, there is sometimes remarkable cognitive dissonance. For example, the French semi-constructivists, Lebesgue, Baire, Borel, voiced skepticism regarding the Axiom of

Choice but implicitly relied on it in some of their important work. More commonly the professional mathematician is simply not interested in foundational or philosophical matters and doesn't regard him/herself as tied to *any* particular, precise construal of mathematical discourse that would imply a position on such matters. The "hermeneutic approach" is a will-o'-the-wisp.

Let me now comment on the changing relationship between practitioners of POM and FOM, on the one hand, and those of mainstream mathematics itself, on the other. From the nineteenth century to at least the 1930's, it was principally mathematicians who undertook foundational studies—from Bolzano, Weierstrass, Dedekind, Cantor, Frege et al. in the nineteenth to the early twentieth with Zermelo with his set theory, Hilbert with his work on geometry and then his proof theory, Brouwer and Heyting with their intuitionistic mathematics, Weyl with his protopredicativism, and then Gödel and Tarski with their great work in metamathematics. In contrast, leading mathematicians in our day for the most part take little interest (and do even less work) in foundations, leaving it principally to logicians (who are in many cases fine mathematicians) and mathematically oriented philosophers. (Category theorists are a special breed, requiring a separate discussion, not possible here.) This is a sorry state of affairs, but we carry on anyway, working on basic issues such as "necessary uses of abstract set theory"(Friedman et al.), the power of predicative mathematics (Feferman, et al), the possibilities of nominalistic reconstructions of (various portions of) mathematics, the nature and significance of various forms of structuralism, and so forth. (While the latter two bodies of work are mainly due to philosophers of mathematics, with respect to structuralism, set theorists and category theorists can claim to have contributed as well, at least indirectly.)

It is an interesting question for historians and social psychologists why the relationship of leading mathematicians to FOM and POM underwent the shift just described. To some extent, no doubt, it had to do with the state of the subjects themselves: in addition to the foundational paradoxes, so much was open to be investigated and formulated that foundational studies must have seemed worth the efforts of early twentieth century mathematicians with a broad and deep knowledge of their subject. Something of a revival occurred in the 1960's with Paul Cohen's seminal work on the continuum problem providing the second half of the independence proof that remained open after Gödel's consistency proof of the mid- '30's; Cohen's work stimulated interest in higher set theory recognized in some leading mathematics departments. A closely related factor may well have been the increasing specialization of disciplines (in the areas of logic and FOM more than in mathematics *per se*, which was already specialized enough). Whether further shifts in attitudes also occurred as part of larger processes is harder to pin down and requires more than my armchair speculations to settle.

One sees some parallels in other fields, e.g. philosophy of physics *vis-à-vis* physics. It was for the most part philosophers that had the patience, in the late twentieth century, to work out detailed modal interpretations of quantum mechanics (more informatively described as "partial hidden variables interpretations") where the main exception having been the Princeton mathematician, Simon Kochen. The point of such interpretations was to solve the measurement problem on a "realist" or "objectivist" basis (by assigning more definite values of quantities than the standard interpretation supports, but not so many as to conflict with various "no hidden variables" theorems). This would improve on the Copenhagen interpretation, with its notorious "cut" between observed system and observer (posing special problems for quantum cosmology which needs to make sense of a quantum state of the whole cosmos). The measurement problem, however—while presentable as a logical contradiction impeccably derived in about three lines from standard assumptions—is not one that most physicists have ever encountered, or else is regarded as a pseudo-problem resulting from some bumbling philosophers' errors (itself a somewhat pardonable error on the part of the physicists, one may admit). On the other hand, the bulk of serious theoretical work on quantum measurement or related foundational problems, starting with Schrödinger, de Broglie, and Einstein, and continuing with Bohm, Everett (author of the so-called "many worlds" interpretation in a doctoral dissertation entitled "Theory of the Universal Wave Function", under John Archibald Wheeler's direction), John S. Bell, and many others, has been carried out by theoretical physicists. Nowadays, they have more license to speculate, it seems, than do philosophers! (If a philosopher had published the original Everett thesis (c.1957), would anyone today even know about it?)

As this cursory sketch indicates, foundational matters were very much on the minds of some of the great founders of quantum mechanics. Not so but for a small minority of their physicist successors. An amusing anecdote will serve as illustration. After a lecture by Nobel laureate Sheldon Glashow (in the '80's at Indiana University), my colleague Doug Hofstader and I had a chance to talk to Glashow at a reception, and we asked him what he thought about the quantum measurement problem. He said emphatically that he simply didn't understand what the problem was, his tone strongly hinting that he regarded it as a pseudo-problem. We protested that of course he must understand it, rehearsing the straightforward derivation of a post-measurement superposition of states of the total object + apparatus system, not definite for any "pointer values" by the standard interpretation, in direct conflict with observation. "Oh, yes, I see that," he replied, "but I don't have to *work* on it, do I?" We cheerfully agreed, remarking that we respected academic freedom!

Since the late 1960's, however, there has been a shift toward greater interest in foundational issues among a number of physicists along with increased communication with philosophers of physics. This is in part traceable to the seminal work of John S. Bell (1964) whose theorem discriminating between quantum mechanics and a very broad class of hidden variables theories led to new and remarkable experiments testing quantum mechanics against the new inequalities (extensions of Bell's original theorem due to Clauser, Shimony, Horne, et al.), with most (and the best) of these confirming quantum predictions. An immediate consequence is that foundational problems associated with quantum measurement and the Einstein-Podolsky-Rosen "paradox" are highlighted as harder than ever to solve. This has informed further new research in physics, such as work of Leggett on whether quantum mechanics governs the "mesoscopic" realm (e.g. collective behavior of particles in superconductivity) or new, ingenious proposals to modify quantum mechanics by introducing spontaneous collapse or localization in dynamical evolution (work of Ghirardi, Rimini, and Weber and Pearl, et al.). All this has provided philosophers of physics much to assess in their pursuit of "experimental metaphysics", as Abner Shimony has aptly called it.

Relatively neglected topics

1. Just how strong must mathematical assumptions be in order to derive the mathematics needed in the sciences (natural and social as well)? The question goes back at least as far as Hilbert, who implied that constructive mathematics in the manner of Brouwer

would deprive the mathematician and scientist of essential tools (law of excluded middle and allied principles) (like depriving a boxer of the use of his fists, according to one famous quote). Hilbert's student, Hermann Weyl, who early on had embraced Brouwer's intuitionism, later came to give it up partly because he doubted the capacity of such truncated mathematics to serve the needs of the sciences (watching with pain as the classical edifice appeared to dissolve before his very eyes, as he put it in one memorable passage in his *Philosophy of Mathematics and Natural Science* (Princeton, 1949, p. 54)). With the remarkable advances in the sciences throughout the twentieth century, especially in physics, with its ever greater uses of abstract mathematics, one might suppose that the skepticism of Hilbert and Weyl regarding the power of constructive mathematics would have been decisively vindicated. Indeed, the rise of the Quinean approach to justifying abstract (classical) infinitistic set theory via its putatively indispensable role in scientific applications might be taken as reflecting lessons of Hilbertian-Weylian skepticism. (After all, if indispensability arguments don't take one beyond constructive number theory and the potentially infinite, why bother with them?)

Surprisingly, however, matters have not turned out that way. On the contrary, with the advent of Bishop's clearer version of constructive analysis (which adds no new, non-classical axioms such as those governing Brouwer's notorious choice sequences, but remains within the confines of intuitionistic logic), large chunks of modern classical functional analysis, including measure theory, have been "constructivized". Classical theorems, if not proved constructively, are approximated via techniques of reformulation, often adding some antecedent hypotheses to the theorems or suitably weakening the conclusion, or both. In addition, the rigorous development of *predicative analysis* by Feferman and others (inspired by Weyl's own work), incorporates even greater swaths of classical mathematics for applications. (In these systems, the logic is classical but sets and functions are affirmed only if they are definable in (countable) mathematical language from the natural numbers or other "already justified" objects in a suitable ordering of extensions of number theory.) With these and related advances in logic and foundations of mathematics, we must now be skeptical of Hilbertian-Weylian skepticism and the Quinean program themselves! (It is as if constructive logic and semi-constructive FOM have caught up with physics.)

Thus, the situation is complicated and the researcher must move with caution, if not trepidation. One needs sufficient background in the natural sciences, especially at least some of its more mathematical portions, in addition to increasingly demanding prerequisites in logic and FOM. But this is not enough. One may prove that a certain version of an applied mathematical theorem is essentially non-constructive, e.g. by the method of weak counterexamples (as I did with a strong version of Gleason's theorem characterizing measures on subspaces of a Hilbert space of dimension > 2), only to learn later that a weaker statement approximating the former version of the theorem *is* constructively provable (as Richman and Bridges, responding to my challenge, showed in the case of Gleason's theorem). In other cases, one may not even contrive a *reductio* via weak counterexample but point to rather large subtheories of demonstrably non-constructive mathematics that appear essential in the statement and proof of an interesting result (as I argued with respect to Hawking-Penrose singularity theorems in the theory of causal structure on semi-Riemannian manifolds). It remains difficult to predict just what can be constructivized; and then it may be a further uncertain matter whether some non-constructive content will actually come to have "scientific application" in the future (e.g. the "preferred basis" component of the strong version of Gleason's theorem that I showed non-constructive). When one moves to the outer reaches of predicative systems of analysis, matters are similarly difficult to assess: one is asking for instances of scientific application which essentially involve the uncountably infinite, e.g. uses of non-separable function spaces (lacking any countable basis). These may be hard to find, and they may exist only on the fringes of theoretical investigation and hence be regarded as indecisive.

For the sake of argument, as well as sound philosophy of science, one should take a rather broad view of "scientific applications of mathematics". Mathematics is used in the sciences not merely to calculate values of functions at given arguments (of course such mathematics is constructive on its face) but to conceptualize and explore a wide variety of idealized models of possible situations. In such a setting, the relevant notion of "justified inference" is not according to rules that preserve constructivity but rather according to rules that preserve truth in models, along classical Tarskian lines. There is no reason why the natural world ought to operate in accordance with the computing capacities of an idealized human mathematician. Bishop said that if God has mathematics to be done, we should let him do it himself. Well, if with Einstein, we move to the "God" of Spinoza, he already *has* done it himself (or, at any rate, a significant portion of it): in the natural world we inhabit! Since evidently he has an inordinate fondness for beetles, who are we to say that he doesn't also have an inordinate fondness for epsilonics and hairy constructive proofs? But we are certainly in no position to say that he does either.

2. Classical analysis depends crucially on the background assumption of infinite totalities, usually taken in the form of the set of natural numbers (or the integers, or the rationals). Of course, it needn't be infinite *sets*; arithmetic and analysis can be presented in a variety of ways, in terms of nominalistically acceptable parts and wholes, second order entities such as concepts or properties, a logic of plurals, and so forth. But somewhere it must be assumed that infinitely many items are available for constructing or recognizing, e.g., convergent sequences of rationals, open and closed intervals, etc. But where does this background assumption of infinitely many things come from and how can it be justified?

Traditional empiricism founders over this question (as Carnap recognized clearly). Classical logicism in the hands of Frege proposed a solution: recognizing first an empty concept and its extension and then a suitably defined immediate successor relation (defined on extensions of readily defined concepts involving just identity, basic logical operations, and already given concepts), one obtains the extension of the concept "bears the ancestral of successor to 0", i.e. the class of Fregean natural numbers, easily seen to be infinite. Unfortunately, this argument relied crucially on unrestricted comprehension, that arbitrary mathematical open sentences determine an extension, which led directly to the Zermelo-Russell paradox. The standardly accepted fix, Zermelo's Separation principle (*Aussonderung*), guarantees only the extension of a concept or open sentence lying within an already given set, and so is useless for deriving an infinite set without blatant circularity. Set theory simply adds its *axiom* of infinity at the outset, setting the problem of justification to one side (and helping provide jobs for philosophers). Other foundational schemes don't fare much better. Russell's type theory resorted to conditionalizing, taking the existence of infinitely many things as a hypothesis of every theorem needing it, thereby sidestepping the problem while maintaining the illusion of a logicist reduction of analysis. Category theory, never very clear about the distinction between axioms as (stipulated) defining conditions on structures (as in abstract algebra) and axioms in the traditional Fregean sense as (asserted) truths about a pre-existing subject matter, offers no improvement, either simply assuming a natural numbers "object" or ignoring the problem entirely. Intuitionism changes the subject, getting by with what an idealized human mathematician can construct—the *potential* infinite—and, in its more radical manifestations, declaring the classical notion of so-called *"actual"* infinity "unintelligible" or "meaningless". Predicativism, to the contrary, embraces classical infinity, usually simply taking the natural numbers as *given*, thereby also failing to address the problem. (Feferman and Hellman's strategy of deriving a natural numbers structure within an elementary theory of finite sets and classes (without finite-set induction), clever though it may be, also sidesteps the basic question of justifying the infinite by assuming, in effect, infinitely many items generated via an ordered pairing function with an urelement under pairing.) The "neo-Fregean " or "neo-logicist" program, far from solving the problem, masks it with its derivation of the Peano axioms from a background second-order logic together with its misnamed "Hume's Principle", treated as somehow "stipulative" ("the number of F = the number of G iff there exists a 1-1 correspondence between the extension of G and that of F" is read as stipulating what it means to talk of numbers), and therefore "justified" if not quite analytic. That this enables a derivation of infinitely many things, and even of an infinite extension, should yield to *modus tollens*: the combination of "Hume's (i.e. Cantor's) principle" and 2d order logic cannot be purely stipulative, precisely because it implies infinity. Nominalistic programs such as Field's may appeal to the coherence or consistency of assumptions about spacetime (e.g. infinite in extent and Archimedean, or infinitely divisible, etc.). But this just amounts to a recasting of the problem in physical-geometric language. How are such assumptions about spacetime any easier to justify than those of Peano arithmetic or Zermelo set theory? Finally, the various structuralisms also end up assuming an axiom of infinity in one form or another.

Perhaps as its author, I ought to say something about modal-structuralism ("MS") in this regard. Using the machinery of mereology and logic of plurals, one can formulate that there are infinitely many individuals. In the MS framework, one then asserts that this is *possible*. While strictly weaker than a claim of infinitely many things *actually* existing, still it is not—at least not obviously—a (modal) *logical* truth. At least there is no contradiction in supposing that every possible world or situation is finite, or so it seems. Especially in view of the *Extendability Principle*, there is no guarantee that a plurality of finite worlds must yield a union of all of them (i.e. of their domains). We are left then appealing to a pragmatic standard of success and utility in practice. The practice includes pure mathematics as well as the empirical sciences. But one would like to be able to say more and with greater decisiveness about so fundamental a matter at the heart of classical mathematics.

Outstanding problems in POM

Just how determinate, semantically, are the key mathematical concepts *finite (infinite)*, *arbitrary subset of a given infinite set*, and *arbitrary ordinal* or *arbitrary set*, and how is determinateness (in whatever degree obtaining) achieved? These three concepts correspond roughly with three ascending levels of logic or theory, namely weak second-order logic, standard full second-order logic, and set theory, say in the form of ZF or ZFC. (The intermediate and highest levels incorporate their predecessors.) Significantly, none of these logics or theories is formally (recursively) axiomatizable, as a consequence of Gödel's Incompleteness Theorem. (What this means in the case of ZF(C) is that no formal system has exactly all set-theoretic truths (even restricted to low infinite levels of "the cumulative hierarchy") as theorems.) That, however, should not be taken to imply that therefore these notions are indeterminate. Indeed, anyone who takes *finite* to be fully determinate (over a given set or totality) (or who takes *natural number*, or *integer*, etc., to be determinate) will uphold this as a counterexample to the claim that formalizability of the relevant logic is a necessary requirement of semantic determinateness.

One might seek determinateness of these purely mathematical concepts by attempting to ground them somehow in the physical world, with the idea that our access to the latter is comprehensible in naturalistic terms so that one would thereby avoid falling back on intuition or any other mysterious faculty or capacity. Hartry Field explored such an approach in connection with the notion of "finite" with mixed results. (See e.g. "Do We Have a Determinate Conception of Finiteness and Natural Number?", in Matthias Schirn, ed., *Philosophy of Mathematics Today* (Oxford University Press 1998), pp. 99-129.) Determinateness turned out to require certain special posits, e.g., on the structure of spacetime (e.g. an Archimedean property). Not only is this problematic in importing what seems an irrelevant contingency into the grounding of pure mathematics (as, e.g., in the case of a postulate of density of space or time), but epistemic circularity can threaten the whole strategy, as rational credibility of such properties of spacetime seems to require prior understanding either of "finite distance" or of quantification over arbitrary wholes of a countable infinity of atomic individuals, as in a mereological statement of induction to describe a concrete model of arithmetic. Similarly, one can try appealing to certain other notions, e.g. in the realm of syntax, say the notion of "sentence of English"; one can point out that the notion of "finite" is incorporated into this, so that if we can understand how the former is determinate, we also thereby have a paradigmatic case of determinateness of the latter. (Further appeal to one-to-one correspondences could then be invoked to extend the concept to other contexts.) But a critic will point out that understanding "sentence of English" depends on prior understanding of "construction [according to certain generative rules] in a *finite* number of steps". All this illustrates how difficult it is to make progress on even the most elementary of the problematic notions listed above.

How can one argue for the determinateness of "arbitrary subset of the natural numbers" or of "all subsets of the natural numbers"? Sharpen the question a bit by making it relative to a determinate totality of the natural numbers (in effect, taking "finite" as determinate). Assuming the notion of "subset of numbers" is perfectly clear, one can appeal to a univocal sense of "all", learned from myriad other uses and contexts. Repetitively, one can insist that this totality of subsets is *maximal*, that it would not be *possible* to add any subsets to it, etc. So what if there is no possibility of enumerating or generating all these subsets (i.e. the infinite ones)? So what if disparate interpretations of "all" could pass undetected by any specific test we might dream up? Isn't such a notion a perfectly good, indeed *strikingly* good, counterexample to Dummettian verificationism in the domain of mathematics? How can one improve on this foot-stamping "argument"? As Burgess pointed out in a critical assessment of Dummett's case for intuitionism ("Dummett's Case for Intuitionism", *Hist. and Phil. of Logic* **5.** 1984:177-194.), that that case is deeply flawed is no substitute for a positive account of meaningfulness of non-constructive infinitistic notions, something we sadly lack.

When it comes to the question of determinateness of notions such as "all ordinals" or "all sets", etc., we are perhaps somewhat further along the way to a resolution, this time in the negative.

Here Dummett's ideas are, in this investigator's opinion, much more on the mark, in particular his description of these notions as inherently "indefinitely extensible", that is, implying methods of transcending any putatively maximal plurality or totality of such objects. Zermelo gave strong voice to this phenomenon toward the end of his remarkable 1930 paper, "*Über Grenzzahlen und Mengenbereiche*", insisting on the "creative progress" of the idea of indefinitely extending a hierarchy of models of his (second-order) ZF axioms (with Choice as a logical principle). What of unbounded quantification over sets or ordinals, which occur all the time in set theory itself? The natural suggestion gleaned from Zermelo's paper is that such quantification is officially to be understood as relativized to a domain (*Mengenbereich*, really, model) of sets, even though the sentence or sentences in question might well be affirmed over some or any extensions of that domain as well. This is reminiscent of Russell's idea of "typical ambiguity". Clearly, one must be careful here not to imply, with one's use of 'some' or 'any', a totality or even plurality of "all possible extensions" of a given model, on pain of contradicting a principle of extendability inherent in "creative progress". Such quantifiers are open-ended: with them we assert something as holding in "*whatever* higher model we ever consider", *without* thereby implying that there is a fixed, maximal plurality of such models. It is similar to the situation in ordinary language: in accepting an unrestricted logical truth, e.g. "Everything is self-identical", we do not commit ourselves even to *making sense* of "absolutely every object" (or even "absolutely every object in Widener Library"). Rather we commit ourselves to applying the self-identity predicate to anything we would ever come to recognize as an object (or an object in Widener Library). Self-identity is conceptually bound up with "object", even if we cannot assign an all-embracing extension to either term. Notice that this open-ended usage is *not* the same as a relativized or restricted usage. Relativization *replaces* open-endedness in specific contexts, but leaves open that some contexts may require open-endedness. In *ordinary* set-theoretic contexts, for example, as Zermelo recognized, relativized quantification works tolerably well. But in *extraordinary* contexts, e.g. that of Zermelo's 1930 paper itself, or at least the philosophical discussion of the final section, relativization would distort the intended meaning of a metaprinciple such as *Extendability*. In *Mathematics without Numbers* (MWON), I formalized several such principles in a modal-logical language (as saying, e.g., that any model of second-order

ZFC that there might be could be mapped isomorphically to an initial segment of one of higher characteristic), at the same time resisting the temptation to give a formal-semantics which might appear to bring in a totality of "all possible ZFC models", which, on the view I proposed, would be self-defeating. This illustrates the advantage of modal logic in FOM: it allows distinctions that are not recognized in non-modal higher-order logic, in this case that between second order (or plural) comprehension "within a world"—allowed without restriction—and comprehension "across worlds"—forbidden in the system of MWON.

We thus see that it is compatible with an objective, broadly "realist" view of mathematics to insist on the intelligibility and determinateness of second-order quantification (i.e. of "all subsets of a given infinite set") while at the same time denying this of putatively unrestricted quantification over "all possible ordinals, sets, strongly inaccessible cardinals, etc.". The cases are, after all, entirely different. In the former, we are *given* a set and then asked to consider all subsets of *it*, or all functions defined on it with values in some other *given* set (e.g. $\{0, 1\}$). That is, these are limited notions, "already restricted" as it were. Moreover, they are commonplace in mathematical practice. (ZFC set theory captures this nicely: the power set of any set exists and has a (Cantor-von Neumann) cardinality (an Aleph indexed by an ordinal), even if we don't know what that cardinality is.) But "all sets" or "all ordinals" in some putatively absolute sense are supposed to be entirely unlimited and unrestricted, and they raise suspicions in much the same way "absolutely all objects" does—or should, at any rate! Not surprisingly, they are quite foreign to ordinary mathematical practice.

13

Jaakko Hintikka

Professor of Philosophy

Boston University, USA

Originally I studied mathematics side by side with philosophy, and I have always been interested in the nature of mathematics, including the nature of mathematical reasoning. But I do not think that I have even thought of myself as embracing any particular philosophy of mathematics. My ideas about mathematics and its foundations have been driven by my work in logic and by the philosophical insights it has inspired.

By this time, such lines of thought have led me to the conclusion that our entire philosophy of mathematics has to be rethought. The reason is that developments in logic have changed or are about to change that subject in a way that put the foundations of mathematics to a new light.

It is not that mathematics itself has not changed. Two hundred years ago mathematics could still have been thought of as a study of number (of different kinds of numbers and functions from numbers to numbers) and space. Beginning with Gauss and Riemann and some of their contemporaries, mathematics has become something much more abstract and general perhaps characterized as the study of different kinds of structures. In this historical perspective the likes of Frege and Russell cannot help appearing strikingly old fashioned, at least when compared with e.g. Hilbert and the best of Husserl. (Some philosophers' question, e.g. whether we need entities called numbers in mathematics are equally antediluvian problems.) Since logic is the natural tool of such study of structures in general, one might expect a collaboration and perhaps even convergence of mathematics and logic. Yet most mathematicians do not think that they need to do logical theory for the purposes of their own work. If foundations are needed, axiomatized set theory is minimally thought of as serving this purpose.

This alienation may well find a justification in the shortcomings of our usual logic. The core area of this logic is known as first-order logic. It is essentially the theory of the existential and universal quantifiers (plus propositional connectives). It is generally thought that the meaning of quantifiers is exhausted by the way they range over a class of values. This overlooks another component of their semantics. The only way we have of expressing on the first-order level the actual objectual dependence and independence of variables of each other is through the formal dependence and independence of the quantifiers of each other to which they are bound. In the received logic, this formal dependence is expressed by the nesting of their scopes. Such nesting has a simple structure (it is e.g. transitive and antisymmetric) and is hence incapable of expressing all possible patterns of dependence and independence.

This defect is corrected in the so-called independence-friendly (IF) logic. (For it, see Hintikka 1996.) It has a better claim to be our unrestricted basic logic than the received first-order logic (RFOL). IF first-order logic is deductively weaker than RFOL and can be thought of as the realization of the intentions of the intuitionists. (For one thing, the law of excluded middle does not hold in it, and consequently the negation it uses is a strong one different from contradictory negation.) However, it is richer than RFOL in its expressive power, and capable of codifying several crucially important mathematical conceptualizations and modes of inference that cannot be handled by RFOL. They include equicardinality, infinity, topological continuity, König's lemma etc.

The strictly construed IF logic can be extended without jeopardizing its elementary character by admitting to it also a sentence-initial contradictory negation.

The properties of IF logic necessitates a re-evaluation of everything in the foundations of mathematics that turns on the nature of logic used in mathematics. For one thing we can now exorcise what is sometimes called "Tarski's Curse": the notion of truth for a (syntactically sufficiently rich) first-order language turns out to be definable in the same language. (See Hintikka 1996, ch. 6 and 2001) This means that virtually everything that has been said of the notion of truth and its definability in mathematics and elsewhere after Tarski's 1935 paper has to be reconsidered.

As was already hinted at, the relation of the intuitions of intuitionists to first-order logic have to be re-examined. Moreover, if IF logic is used as the logic of a system of elementary number theory, the (model-theoretical) consistency of the system turns out to be provable in the same system. Hence Hilbert's foundational project and its would-be refutation by Gödel are put to an entirely new light.

Another topic that is deeply affected is the so-called axiom of choice. It turns out to be a truth of IF logic, not a separate set-theoretical assumption, as it is usually taken to be. Indeed, an intuitively obvious reformulation of RFOL already turns the "axiom" of choice into a first-order logical truth. (See Hintikka, forthcoming.). Thus the status of the "axiom" of choice in mathematics and its philosophy has to be re-evaluated, including Hilbert's project of reconstituting the "axiom" on the first-order level by means of his epsilon technique.

But this is not the end of the story. Even though much more can be done in an (extended) IF logic, it does not capture all the modes of reasoning needed in mathematics any more than RFOL does. What more is needed? The two main candidates are higher-order logic and set theory. The former was the choice of Frege (1884) as well as of Russell and Whitehead (1910-13). Recently, this option has fallen into something of disrepute. Working mathematicians find set theory more convenient, and many philosophers follow Quine and branded higher-order logic as "set theory in sheep's clothing". But the main reason for this preference is fallacious. It is the absence of complete axiomatization of higher-order logic. But this is a little more than a push on the word "axiomatization". Rightly understood, the axiomatization of the truths of some one field means representing them as logical consequences of a system of axioms. This relation is completely formal, depending as it does only on the forms of the propositions involved. But in the confused thinking of many philosophers it is also required that relations of logical consequence must be exhausted by mechanical rules of inference. Whether this can be done in different cases is an interesting question, but its interest is a computational one, not a philosophical one. It does not deal with the limits of what logic can do or what a mathematician can do. It shows a limitation of what computers can do. For instance, Gödel's first incompleteness theorem (Gödel 1931) has no relevance to whether arithmetic can be axiomatized in the more general sense. As a consequence it does not show anything about what logical conceptualizations and logical reasoning can or cannot do in mathematics. It only shows that a computer cannot be programmed to spew out all arithmetical truths one after the other. I once tried to illustrate the resulting need of discovering even more and more logical truths by saying that a logician is like a housewife: her work is never done. But this is a wrong metaphor because much of a housewife's work can be mechanized, which is precisely what cannot be done in logic. Hence higher-order logics (or IF logic, for that matter) cannot be faulted because they are not axiomatizable in the sense of mechanizable.

As to the latter candidate, set theory in its present shape has turned out to be the wrong choice as a medium of mathematical reasoning. For one thing, set theory in its usual incarnation suffers from a terminal case of schizophrenia. It is in the first-place a study of certain infinite structures. But it also pretends to be a depository of the modes of inference needed in mathematics. These two tasks cannot both be accomplished by the same theory. In that theory, a logician already needs some of the modes of reasoning that the set theory is supposed to capture. If you somehow try to codify those modes of inference in the properties of the models of set theory, you still need some logic for the purpose of studying these models for the purpose of bringing out the consequences of those principles of reasoning. The modes of reasoning codified in the usual axiomatizations of set theory do not serve this purpose adequately. What used to be called semantical paradoxes of set theory are symptoms of this failure, even though they do not amount to formal contradictions.

The usual form of set theory in our day and age is a first-order axiomatic theory using the received first-order logic. Such a set theory is useless as a foundational project. Why? In order to see the reason, consider a model M of a first-order axiomatic set theory. This model can be considered in two ways. You can interpret in it the membership relation $\in$ like any other two-place relation. But you can also require that in M or in some part of M $\in$ is indeed interpreted as if the membership relation. This imposes a very special structure M considered as a first-order model. For one thing, for certain classes of individuals in M there must be an individual to which they all, and only they, bear this relation.

It is known that the entire model M can never be interpreted in this way set-theoretically. But it is hoped and prayed that these parts of M for which this cannot be done are so outlandish (perhaps consisting only of very large sets) that are not relevant to the pursuits of a working mathematician and that by introducing new axioms we can make our models more amenable to a set theoretical interpretation.

If this was the motivation of axiomatic set theorists, their prayers were not heard. There are in the usual first-order axiomatic set theories theorems that are false in their set-theoretical interpretation and yet do not have anything to do with very large sets. This falsity cannot be removed by the introduction of any new axioms. (Admittedly, at this time, none of the known false theorems have normal mathematical interest. But this means that we are in the same position as logicians before. In the same way as Gödel's first incompleteness result received mathematically interesting instances in the work of Paris and Harrington 1977.) The existence of such false theorems implies that results concerning the possibility or impossibility of a proposition in the usual first-order axiom systems of set theory have no direct relevance to the truth or falsity of the proposition. Ergo, Gödel's and Paul Cohen's unprovability results have as such little mathematical interest.(Cf. Gödel 1940,Cohen 1966). As a further consequence, it is a rank mistake to claim that the continuum problem has been "solved" by such results. (Gödel did not make this mistake; see his paper in Benacerraf and Putnam 1983.) Hence, all told, first-order axiomatic set theory is a misleading guide to set-theoretical truth.

It is also an inefficient guide. For, as was pointed out, the foundational consequences of axiomatic set theory have to be read from their models. But axiomatic set theory is a very poor model theory of set theoretical structures. In such a model theory, an important role is sometimes played by sets that are not definable in axiomatic set theory. Such sets cannot be utilized in axiomatic set theory used as its own model theory. Hence it is only to be expected that all sorts of set-theoretical truths are impossible to prove in first-order axiomatic set theory.

Hence no mathematical, foundational or philosophical theories can be based on what happens in first-order axiomatic set theories.

And there is an even more striking foundational objection to the use of set theory: It is dispensable. Set theory was resorted to in the foundations of mathematics because ordinary first-order logic did not capture all the modes of reasoning used in mathematics. More explicitly, it does not capture the modes of inference that depend on the nature of higher-order entities. Actually, second-order logic gives us all we need in the foundations of mathematics.

Now IF logic is much weaker than second-order logic. In fact it is equivalent to the Σ_1^1-fragment of second-order logic. Adding a sentence-initial contradictory negation only brings in the Π_1^1 fragment. But what happens if we allow the contradictory negation to occur also within the syntactical scope of quantifiers? (See here

Hintikka 2006.) We need then a much stronger semantics than the usual game-theoretical semantics for IF logic. We need applications of the law of excluded middle to increasingly complex sets. They transcend widely the elementary character of IF logic. However, we are still moving on the first-order level, and these applications of tertium non datur are all we need to make sense of contradictory negation anywhere in formulas that are IF otherwise. The striking fact is that the resulting first-order logic is as strong as the entire second-order logic. This result shows in effect that all of mathematical reasoning can be carried out on the first-order level. We do not have to search for answers to any questions about higher-order entities, including the conditions of their existence. In principle, all we need to consider are configurations of particular individuals. If my last name were Kronecker, I would perhaps say that God created particular objects and the different structures they can form, everything else is *Menschenwerk*. And I suspect that this claim is what Hilbert's misnamed "formalism" at bottom amounted to.

Set theory will continue to exist as a study of certain infinite structures. But it should never again be considered as a codification of our modes of reasoning about higher-order entities. More generally, no foundational or philosophical theorizing must even be based on the received axiomatic set theories.

These remarks leave untouched what some people seem to think as the most conspicuous recent development in logic. This development is the rise of even so many different so-called non-classical logics, including logics of nonmonotonic reasoning. (For such logics, see e.g. Gabbay et al., 1994.) What is their philosophical relevance? (See here the postscript to Hintikka 2006.) Before we can even ask this question, we have to know what the intended interpretation of their sundry formalisms is. In most cases, this interpretation is left almost completely in the dark. However, in some of the most interesting representative cases, it is possible to understand or at least to interpret what is involved. These logics are enthymemic logics. They are systematizations of inferences from premises that are partly unexpressed. For instance, circumscriptive inferences rely on the assumption that the information given in the initial premises is complete. Inductive inferences rely (as Carnap's work in effect shows) on assumptions concerning the orderliness of the universe of discourse. (See Carnap 1952, and note that his lambda can be interpreted as a conjecture concerning the orderliness of one's universe.)

Such inferences can be intriguing and otherwise worth studying. But from a general theoretical viewpoint such logics are *Ersetz* theories. What should be done is to spell out and make explicitly expressible the tacit premises of the enthymemic inferences. Only when this is done can we as much as to decide whether any really new logical principles are needed. At least, so far I have not seen any real evidence that theoretically interesting novelties are needed, let alone already discovered.

Before the dust settles and the significance of these different developments is made clearer, traditional discussions in the philosophy of mathematics run the risk of being hopelessly out of date.

References

Benacerraf, Paul and Hilary Putnam, 1983, editors, *Philosophy of Mathematics*, 2nd edition, Cambridge University Press.

Carnap, Rudolf, 1952, *The Continuum of Inductive Methods*, University of Chicago Press.

Cohen, Paul, 1966, *Set Theory and the Continuum Hypothesis*, W. A. Benjamin, New York and Amsterdam.

Frege, G., 1884, *The Foundations of Arithmetic*, original German as *Die Grundlagen der Arithmetik, Eine logisch mathematische Unterschung über den Begriff der Zahl*, Breslau; English translation, translated by J. L. Austin, Northwestern University Press, 1980.

Gabbay, Dov M., C. J. Hogger, and J. A. Robinson, 1994, editors, *Handbook of Logic in Artificial Intelligence and Logic Programming, vol. 3, Non-monotonic Reasoning*, Oxford University Press.

Gödel, Kurt, 1931, "On formally undecidable propositions of *Principia Mathematica* and other related systems", original German in *Monatshefte für Mathematik und Physik*, vol 38, pp. 173–198; English translation in Jean van Heijenoort, editor, *From Frege to Gödel*, Cambridge, MA, Harvard University Press, 1967, pp. 596–616.

Gödel, Kurt, 1940, "The Consistency of the Axiom of Choice and of the Continuum-Hypothesis with the Axioms of Set Theory", *Annals of Mathematics Studies*, vol. 3, Princeton University Press, Princeton.

Hintikka, Jaakko, 1996, *Principles of Mathematics Revisited*, Cambridge University Press.

Hintikka, Jaakko, 2001, "Post-Tarskian truth", *Synthese*, vol. 126, pp. 17–36.

Hintikka, Jaakko, 2004, "Independence-friendly logic and axiomatic set theory", *Annals of Pure and Applied Logic*, vol. 126, pp. 313–333.

Hintikka, Jaakko, 2006, "Truth, negation and other basic notions of logic", in van Benthem et al., editors, *The Age of Alternative Logics*, Springer, pp. 195–219.

Hintikka, Jaakko, forthcoming, "Truth, axiom of choice, and axiomatic set theory".

Hintikka, Jaakko, and Besim Karakadilar, 2006, "How to prove the consistency of arithmetic", *Acta Philophica Fennica*, vol. 78, pp. 1–15.

Paris, Jeff and Leo Harrington, 1977, "A mathematical incompleteness in Peano Arithmetic", in John Barwise, editor, *Handbook of Mathematical Logic*, North-Holland.

Quine, Willard V. O, 1970, *Philosophy of Logic*, Prentice-Hall.

Tarski, Alfred, 1935, "Der Wahrheitsbegriff in den formalisierten Sprachen", *Studia Philosophica*, vol. 1, pp. 261-405; English translation in A. Tarski, *Logic, Semantics, and Metamathematics*, Clarendon Press, Oxford, 1956, pp. 152–277.

Whitehead, Alfred, N., and Bertrand Russell, 1910-13, *Principia Mathematica* (2 volumes), Cambridge University Press.

14

Thomas Jech

Professor of Mathematics, Emeritus
The Pennsylvania State University, USA
Senior Research Scientist, Mathematical Institute
The Czech Academy of Sciences, Czech Republic

Why were you initially drawn to the foundations of mathematics and/or the philosophy of mathematics?

In 1963, I joined a group of young mathematicians led by Petr Vopenka of Charles University to work in Set Theory and Foundations of Mathematics. When I read Gödel's paper on the axiom of choice and the continuum hypothesis, I was hooked for life.

What examples from your work (or the work of others) illustrate the use of mathematics for philosophy?

Most of my work has dealt with forcing and large cardinals, influenced by Paul Cohen, Robert Solovay and others. This area of mathematics helps to understand the basic principles underlying mathematical reasoning.

What is the proper role of philosophy of mathematics in relation to logic, foundations of mathematics, the traditional core areas of mathematics, and science?

Philosophy of mathematics can be helpful in explaining the significance of technical progress in mathematics. It is essential though that this is done by practicing mathematicians who have the sufficient insight.

14. Thomas Jech

What are the most important open problems in the philosophy of mathematics and what are the prospects for progress?

As a corollary to the above, the open problems emerge from the ongoing progress in mathematics.

15
H. Jerome Keisler

Vilas Professor of Mathematics, Emeritus

University of Wisconsin, Madison, USA

Why were you initially drawn to the foundations of mathematics and/or the philosophy of mathematics?

I was initially drawn to mathematical logic when I discovered the book of Hilbert and Ackermann in the summer of 1956, after my first undergraduate year at CalTech. I was attracted by the fundamental nature of the subject, and by the prospect of reaching the frontier of research quickly. At the time, logic had an air of mystery for me because the field was small and there were no specialists on the faculty at Caltech.

In the fall of 1956 C.C.Chang (then at USC and later at UCLA) attended the algebra seminar at Caltech as a visitor. He became my primary mentor in my undergraduate years, when I got my start in research in model theory and produced two papers, and was later to be my co-author for two books and several articles. With ample warning, C.C. prepared me for the crucible of graduate study at Berkeley (1959–1961), which was aptly described in the wonderful book *Alfred Tarski. Life and Logic* by Anita and Solomon Feferman.

What examples from your work (or the work of others) illustrate the use of mathematics for philosophy?

The paper "From accessible to inaccessible cardinals" [1964] with A. Tarski is the starting point for the systematic study of large cardinals and strong axioms of infinity, which has since become a central theme in set theory and the foundations of mathematics. The paper gave evidence that the notions of weakly compact cardinals, measurable cardinals, and compact cardinals are natural analogues of the notion of an infinite cardinal which arise from a wide variety of mathematical problems. More recently, stronger axioms of infinity, such as the existence of many Woodin cardinals, have gained prominence.

The book *Model Theory* [1973] with C.C.Chang organized and became the primary reference for the subject that provides the fundamental justification of formal methods by providing the link between the syntax of a formal language and its meaning. Since the book was written, model theory has split into many branches, each with its own methods, such as the model theory of tame structures, finite model theory, models of set theory and arithmetic, nonstandard analysis, and model theories for a variety of non-first order logics and modal logics. Our book is now the trunk of the tree which supports all these branches. The book also has the following real-life illustration of the Russell paradox: "we dedicate our book to all model theorists who have never dedicated a book to themselves."

The model theory paradigm which was originally developed for first order logic in the first half of the 20th century is of central importance in many other settings. When investigating an area by formal methods, one often starts by developing a formal language suited for the area of interest, and then developing a model theory for the formal language. This general approach is illustrated in my monographs *Continuous Model Theory* [1966] with C. C. Chang, *Model Theory for Infinitary Logic* [1971], and *Model Theory of Stochastic Processes* with S. Fajardo [2002].

In my research I have been attracted to A. Robinson's nonstandard analysis, a formal approach that has been underused and captures some powerful intuitive ideas, such as hyperfinite sets and infinitesimals, which can help in the discovery of new concepts and results. (The importance of mathematical intuition in general will be discussed later in my answer to the next question.) As the history of calculus suggests, the intuitive idea of an infinitesimal can be grasped at an elementary level, and makes it easier for beginning calculus students to absorb and use the basic concepts of limit, derivative, and integral. To make the infinitesimal approach available to beginners, I wrote the book *Elementary Calculus, an Approach Using Infinitesimals*, (1976, 1986, open source online edition [2000]). I think of this as planting a seed, with the hope that early exposure to the intuitive idea of an infinitesimal will have long term benefits.

In the paper "The hyperreal line" [1994], I discussed some of the issues in the philosophy of mathematics that are raised by the concept of an infinitesimal. The series of papers [1984], [1986], [2006a], [2006b] on the proof-theoretic strength of nonstandard analysis is motivated by the philosophical question of how and where one can expect nonstandard methods to lead to new mathematical results.

In the paper "An impossibility theorem on belief in games" [2006a] with A. Brandenburger, we use formal methods to analyze the following new paradox on iterated beliefs: "Ann believes that Bob assumes that Ann believes that Bob's assumption is wrong".

What is the proper role of philosophy of mathematics in relation to logic, foundations of mathematics, the traditional core areas of mathematics, and science?

As a mathematician, I will write about the proper role of mathematics, and in particular of formal methods in mathematics, in relation to other disciplines. In order to answer this question, it seems necessary to take a stance on the question of mathematical existence. According to the classification in P. Maddy's recent paper "Mathematical Existence" [M], my position is somewhere between arealism and thin realism. Briefly, mathematical intuition about infinite objects exists, but I reserve judgment on whether or not the infinite objects themselves exist.

Gödel [G] wrote in 1964 that "the question of the objective existence of the objects of mathematical intuition ... is an exact replica of the question of the objective existence of the outer world." "The mere psychological fact of the existence of an intuition which is sufficiently clear to produce the axioms of set theory and an open series of extensions of them suffices to give meaning to the question of the truth or falsity of propositions like Cantor's continuum hypothesis."

I agree with Gödel that a mathematical intuition exists which is sufficiently clear to produce the axioms of set theory and an open series of extensions of them. I also agree that mathematical intuitions exist independently of the observer. However, I reserve judgment on whether the intuition is or can be clear enough to give meaning to the question of the truth or falsity of propositions like Cantor's continuum hypothesis.

I view mathematical research as exploring mathematical intuitions. This is ordinarily done by a community of mathematicians collectively building a formal system and proving a large number of theorems. Formal systems are used to clarify, sharpen, and communicate intuitive observations. We use mathematical intuition to find new axioms, make conjectures, and find proofs of theorems which follow from the axioms. This provides evidence of the existence of mathematical intuition. As Gödel pointed out, the consequences of the axioms, in turn, can be checked against intuition, and provide evidence for or against the axioms, in a manner analogous to the scientific method of using observations to test a theory. Computers give mathematicians another tool for testing axioms, as well as for experimentation which can stimulate mathematical intuition.

Mathematical intuition occur in a variety of settings, for example: the cumulative hierarchy in set theory; constructive mathematics; computational complexity; modal logic; nonstandard universes; category theory; probability theory; string theory; biological and social sciences. One should expect that completely different intuitive viewpoints are possible, and that some of them will be discovered in the future.

A successful example which is often cited is the use of the intuitive concept of the cumulative hierarchy of sets to find the basic axioms of set theory, add axioms of infinity, and draw conclusions about sets at the lower levels. One can formally represent most of mathematics within set theory. However, some caution is needed because it is often hard to transfer mathematical intuition from one setting to another, or even from one subarea of mathematics to another. For instance, when one is immersed in the usual set-theoretic hierarchy, it is hard to think intuitively in terms of functors in category theory, or hyperfinite sets in a nonstandard universe, or constructive existence. For this reason, I am in favor of a pluralistic approach which encourages the unrestricted exploration of mathematical intuition in different settings. Fruitful ideas can be discovered in one setting but obscured in others.

Robinson's nonstandard analysis is one of many examples where different mathematical intuitions have led to new concepts and results. Constructive mathematics is another example of this kind. It is a good idea to find out what can be done by constructive methods, but a bad idea to limit mathematics to such methods.

I see two roles for mathematics in relation to other disciplines. One role is to provide algorithms to solve practical problems. A second, more interesting role, where mathematical intuition comes into play, is to create models or theories which shed light on some phenomenon that one wishes to understand. From this standpoint, pure mathematics can be viewed as the study of a mathematical intuition for its own sake, rather than for some outside phenomenon.

One can get some insight into the role of mathematics in relation to other disciplines by looking at a particular area. For example, in the area of mathematical economics there is a very substantial literature on exchange economies. A real-life exchange economy has a large but finite set of small agents who interact in some way, and one would like to understand phenomena such as coalitions, equilibria, the movement of prices, and behavior under uncertainty. In the literature, three different ways of representing exchange economies have been developed and used extensively. These are: an increasing sequence of finite economies where one studies the asymptotic behavior, an economy with a continuum of agents, and an economy with a hyperfinite set of infinitesimal agents. Even though one knows that there are only finitely many agents in a real-life economy, many phenomena are best understood intuitively by using infinite structures. Moreover, the three different approaches involve different mathematical intuitions which are helpful in understanding different phenomena.

What is going on here is that well-behaved infinite objects are intuitively simpler than large complicated finite objects. In all real situations (even in physics) one wishes to explain a large but finite set of observations about the world. The role of mathematics is to use mathematical intuition, often about infinite objects, to help understand the observations. One does not try to give an exact description, but instead searches for an idea which is simple enough to be grasped intuitively and which somehow captures the essential features of the phenomenon. For this reason, mathematics can serve its role best if one has the flexibility to exploit different mathematical intuitions.

What do you consider the most neglected topics and/or contributions in late 20th century philosophy of mathematics?

What are the most important open problems in the philosophy of mathematics and what are the prospects for progress?

I begin with a disclaimer. The problems which I mention are not new, have not been neglected at all, and have been discussed at length in the literature with a great deal of speculation.

The problem which underlies the whole discussion above is:

A. What is mathematical intuition?

I believe that this problem is closely related to the following two problems:

B. Do infinite mathematical objects exist?

C. How does the brain work?

I suspect that further progress on question A will depend on progress on question C. One would expect a great deal of progress in the future on how the brain works, more likely by biologists rather than mathematicians or philosophers. In order to make progress on question A, a person would need both experience with mathematical intuition, and knowledge yet to be discovered about how the brain works.

As for question B, we have only finitely many observations of the real world at our disposal, and these give us no way to tell whether or not infinite objects actually exist. There is also the possibility that infinite objects exist but do not fit any intuitive viewpoint. I have distinguished between mathematical intuition about infinite objects and the objects themselves. Perhaps a better understanding of mathematical intuition about infinite objects will tell us whether this distinction is tenable.

I will close by mentioning two other questions which I consider to be interesting and important, but see little hope for progress in the near future.

D. What would happen if a contradiction were found in a very weak system on which we rely, such as primitive recursive arithmetic?

E. Can there be an experiment which establishes the physical existence of an infinite object? If not, can one prove that there cannot be such an experiment?

References

[1964] H. J. Keisler and A. Tarski. "From accessible to inaccessible cardinals." *Fund. Math.* **53** (1964), pp. 225–308.

[1966] C. C. Chang and H. J. Keisler. *Continuous Model Theory.* Annals of Math. Studies **58** (1966), xii+165 pages.

[1971] H. J. Keisler. *Model Theory for Infinitary Logic* (1971), North-Holland 1971, x+208 pages.

[1973] C. C. Chang and H. J. Keisler. *Model Theory* (Second edition 1977, Third edition 1990), North-Holland (1973), 554 pages.

[1984] C. W. Henson, M. Kaufmann, and H. J. Keisler. "The strength of nonstandard methods in arithmetic," *J. Symb. Logic*, **49** (1984), pp. 1039–1058.

[1986] C. W. Henson and H. J. Keisler. "On the strength of nonstandard analysis," *J. Symb.Logic*, **51** (1986), pp. 377–386.

[1994] H. J. Keisler. "The hyppereal line." *In Real Numbers, Generalizations of the Reals, and Theories of Continua,*" ed. by P. Erlich, Kluwer Academic Publishers, pp. 207–237, 1994.

[2000] H. J. Keisler. *Elementary Calculus, An Approach Using Infinitesimals*, xviii+940 pages. (First edition, Prindle, Weber and Schmidt 1976, Second Edition 1986, Online Edition with Creative Commons License 2000).

[2002] S. Fajardo and H. J. Keisler. *Model Theory of Stochastic Processes*. Lecture Notes in Logic, Association for Symbolic Logic (2002), xii+136 pages.

[2006a] A. Brandenburger and H. J. Keisler. "An impossibility theorem for beliefs in games," *Studia Logica* **84** (2006), pp. 211–240.

[2006b] H. J. Keisler. "Nonstandard arithmetic and reverse mathematics," *Bulletin of Symbolic Logic* **12** (2006), pp. 100–125.

[2007] H. J. Keisler. "The Strength of Nonstandard Analysis." Pages 3–26 in *The Strength of Nonstandard Analysis*, edited by I. Van den Berg and V. Neves (2007).

[G] K. Gödel. "What is Cantor's continuum problem?" *In Philosophy of Mathematics*, Selected Readings, edited by P. Benacerraf and H. Putnam, Prentice-Hall 1964.

[M] P. Maddy. "Mathematical existence." *Bulletin of Symbolic Logic* **11** (2005), pp. 351–376.

16
Ulrich Kohlenbach

Professor of Mathematics
Technische Universität Darmstadt
Darmstadt, Germany

Why were you initially drawn to the foundations of mathematics and/or the philosophy of mathematics?

At the age of 13 or so some initial interest in philosophy and Aristotelian logic was prompted by my classes in Ancient Greek language which was a main emphasis of study at my high school.

My real interest in the foundations of mathematics, however, started at the age of 17 during my last year at high school. Our mathematics teacher had the idea to have each of us to write an extended essay on some period in the history of mathematics. He designed a list of 20 topics starting from ancient mathematics to the beginning 20th century. The very day the topics could be chosen I was ill and could not attend school. When I finally was back in school I had to learn that only topic no. 20 on 'Cantor, Dedekind, Hilbert' was left, apparently because everybody had figured out that a topic touching on comparatively recent mathematics would be more difficult to deal with than, say, Babylonian mathematics. After I had overcome some initial shock I went to the university library in Frankfurt to get hold of the collected works of G. Cantor, D. Hilbert as well as R. Dedekind's 'Was sind und was sollen die Zahlen' and some popular treatments of the 'foundational crisis' at the early 20's century. Immediately, I got excited about the topic. After having finished the essay I was determined to study philosophy and mathematics with the aim to become a logician. During my first 1-2 years at the University of Frankfurt I focused on philosophy but I soon realized that in order to become able to prove new results in logic (rather than just discussing the early history of mathematical logic) I would need a solid background in mathematics. My teacher in Mathematical

Logic at Frankfurt was Professor H. Luckhardt who pointed me towards the writings of A.S. Troelstra (most notably [19]). Via Luckhardt I also got under the influence of G. Kreisel's views on logic and, in particular, proof theory (see e.g. [15, 16]). Much of my subsequent work has been stimulated by Kreisel's ideas on the unwinding of proofs in mathematics and the necessity to pay for the latter some 'entrance fee' by learning enough core mathematics. So I was glad to get during my PhD studies an assistant position in the group 'Real Analysis and Potential Theory' with Professor J. Bliedtner and after that in 'Analytic Number Theory and Functional Analysis' with Professor W. Schwarz.

What examples from your work (or the work of others) illustrate the use of mathematics for philosophy?

Well, as an ancient use of mathematics in philosophy one may point to the refutation of the Pythagorean philosophy by the proof of the irrationality of the length of the diagonal in the unit square.

From my work I can only address possible uses of mathematical logic for the philosophy of mathematics: while in my area originally pre-formal concepts from the philosophy of mathematics have been instrumental to guide certain mathematical developments (see the next question), applications in the opposite direction in my view mainly consist in correcting errors in some philosophically inclined foundational debates. This e.g. concerns the topic of constructive foundations of mathematics. A common view is that a (fully) constructive proof per se is superior (contains more information) than a prima facie classical argument. Experience from the unwinding of classical arguments, however, shows that usually only small parts of a proof contribute to the computational content of a proof while the bulk of the proof may well be proven ineffectively. Sometimes, using a more ineffective argument helps to push further parts of a proof into lemmas which are not needed to be analyzed at all. Replacing ineffective analytic or geometric steps in a proof by constructive induction arguments can result in bounds that are much less good than those extractable from the original classical proofs (see e.g. [7]). Often even ineffective parts in a proof which **do** matter for the computational content can be systematically transformed in such a way that computationally relevant information can be read off. E.g. this can be done by suitable forms of proof interpretations, notably of Gödel's functional ('Dialectica') interpretation ([14]).

In this sense already the original proof can be seen as **implicitly** containing additional computational content even in cases where the latter seems absent at first.

As a justification for a global approach to constructivity (based on intuitionistic logic) often useful properties of the resulting constructive formal systems are pointed to: e.g. usually systems of intuitionistic analysis satisfy the so-called fan rule which, for instance, allows one to infer uniform convergence from pointwise convergence on compact spaces. However, it can be shown that this property also holds for such systems if one adds the binary ('weak') König's lemma WKL or even König's lemma KL (and - since we can include the axiom of choice schema - also the uniform versions of these principles) as well as the Markov principle (in the absence of the Markov principle even full comprehension in all types for all negated formulas may be added, [8, 9, 12]). This is remarkable since it was the very intuitionistic rejection of WKL and related principles which led to the formulation of the fan principle in intuitionistic mathematics. Hence if the motivation for finding a constructive proof is to be able to use the fan rule then there is no reason not to allow e.g. WKL in such a proof despite of the fact that this principle is not effective as long as the proof is constructive relative to WKL. If the matrix in the instance of the fan rule at hand is only purely existential than even full classical logic can be allowed and one still has the closure under the corresponding instance of the rule ([5, 7]). In fact, in my work I have shown that e.g. proofs of uniqueness theorems in best approximation theory that are based on WKL carry just as much computational content (namely a so-called modulus of uniqueness or 'rate of strong unicity') as fully constructive proofs (even of strong positive reformulations of) uniqueness (see [5, 6, 13]).

In connection with the common misconception that a constructive proof always is better than a classical argument it has been overlooked that occasionally it pays off to view a constructive proof as a classical one which first needs a constructivization via negative translation: consider an implication $A \to B$ where A is a $\forall \exists \forall$-('Π_3^0')-sentence while B a $\forall \exists$-('Π_2^0')-sentence and suppose that $A \to B$ is provable in intuitionistic ('Heyting') arithmetic HA. The straightforward constructive interpretation of (a constructive proof of) such an implication (as spelled out e.g. by Kreisel's modified realizability interpretation) is that it provides a procedure F that transforms any witnessing ('Skolem') function f for the premise A into a witnessing functions g for the conclusion. This interpretation, however, is very weak in cases where A does not have an effective witnessing function f. If one views that proof of $A \to B$ instead as a classical proof and applies negative translation to it, then the situation improves: both in the premise A as well as in the conclusion B the existential quantifiers get weakened by a double negation in front. However, this weakening of the premise is much more severe than that of the conclusion. From the latter one easily recovers B using the aforementioned Markov principle. Using now instead of modified realizability a technique supporting the Markov principle, namely Gödel's functional interpretation, one obtains a witness function for the conclusion in a witness of the so-called no-counterexample interpretation ([15]) of the premise A which is a much weaker (usually subrecursive of low complexity) input than the (in general noncomputable) Skolem function seemingly required by the original constructive reading of the implication.

In the context of predicative foundations of mathematics ([20]) special emphasis has been given to the fact that most function spaces used in scientifically applicable functional analysis are separable and so to a large extent can be treated based on arithmetical comprehension ([2]). Similarly, in the program of so-called reverse mathematics (initiated H. Friedman and S. Simpson,[18]) one works – following Hilbert and Bernays ([4]) – in systems (fragments of 2nd order arithmetic) whose very language is so restricted that only separable spaces can be represented (Hilbert and Bernays actually do allow 3rd order parameters). Similar restrictions (to separable spaces) are put forward in constructive mathematics (e.g. Bishop's constructive analysis [1]) as an instance of the general rule of avoiding 'pseudo-generality' and also play an important role in intuitionistic analysis via so-called standard representation of complete separable ('Polish') metric spaces. Recent work in the logical analysis of proofs, however, has shown that even if one is interested primarily in separable spaces it can be crucial to observe that a given proof does not use separability: general logical metatheorems ([10]) guarantee the extractability of uniform bounds that are independent not only from parameters in compact spaces but from parameters in metrically bounded spaces (as well as of self-mappings of such spaces) as long as only general facts about the class of the spaces in question are used which have a strong uniformity built-in. This applies to metric, hyperbolic, CAT(0), normed, uniformly convex and inner product spaces among many others. However, the uniform version of separability translates into total boundedness of metrically bounded subspaces and hence – modulo completeness – compactness. So in order to be able to diagnose uniformity in the absence of compactness it is critical to observe that separability is not used. For much refined recent metatheorem which does not even require the boundedness of the whole underlying space but only bounds on local distances see [3]. A survey on the numerous applications of this approach in metric fixed point theory is given in [11].

So while it **is** of foundational interest that large parts of mathematics can be formalized in weak predicative systems due to the separability of most spaces of interest it would, nevertheless, be philosophically misguided to restrict ones attention in the philosophy of mathematics to those parts only.

What is the proper role of philosophy of mathematics in relation to logic, foundations of mathematics, the traditional core areas of mathematics, and science?

One main direction of influence from the philosophy of mathematics to the foundations of logic and mathematical logic has been the result of bringing concepts first developed in philosophy (not only of mathematics) into a formal context and deriving clear cut mathematical consequences from this transition. A particularly striking example is the Liar paradox which, when phrased in different formal contexts, has played an enormous role in the development of mathematical logic:

1. When stated in a formalized language, say of Peano arithmetic PA, it implies the undefinability of a truth predicate for PA in the language of PA (A. Tarski).

2. When applied to the proof predicate for PA it yields Gödel's first incompleteness theorem.

3. When truth is replaced by 'terminates' it yields the undecidability of the halting problem (A. Turing, A. Church).

Another example of a deep mathematical use of a concept from the philosophy of the mathematics is the notion of predicative definability and its (impredicative) use by Gödel to construct the constructible hierarchy and thereby the (relative) consistency of the continuum hypothesis.

Leibniz' idea of possible worlds led to the development of semantics for modal and intuitionistic logic and so, in a sense, to the notion of forcing (discovered originally by P. Cohen). The latter also has some similarities with the concept of the development of mathematical knowledge in stages as formulated in Brouwer's theory of the 'creative subject'.

Different formalizations of the informal (so-called Brouwer-Heyting-Kolmogorov) semantics of intuitionistic logic have led to numerous important technical devices such as realizability and functional interpretations and play (in the form of the 'proofs-as-programs' paradigm) an increasing role in computer science.

What are the most important open problems in the philosophy of mathematics and what are the prospects for progress?

In my view one of the most important problems in the philosophy of mathematics as well as the foundations of mathematics is to explain why in actually existing core mathematics apparently only a tiny part (both in terms of proof theoretic strength as well as w.r.t. the logical complexity involved) of the vast resources provided by formal systems such as Zermelo Fraenkel set theory is exploited. This concerns the facts that (i) only weak set existence axioms are used and (ii) only formulas of low quantifier complexity show up in ordinary mathematics: despite of the incompleteness of PA usually small fragments of PA suffice to proof statements that can be expressed in the language of PA. Reverse mathematics has shown that large parts of mathematics can be carried out in systems conservative over primitive recursive arithmetics PRA ([17]) and even larger parts in systems conservative over PA ([2]). Although proofs in logic in general have underlying Herbrand disjunctions whose length is nonelementary in the basic proof data (R. Statman, V. Orevkov), in practice such disjunctions usually are rather simple and short. The question now is whether these empirical observations reflect a general mathematical fact, namely that proof theoretic strength and logical complexity have no real relevance for mathematics, or whether it indicates that there are relevant parts of (future) mathematics which simply have been overlooked just because they require a better understanding of strong set theoretic concepts and statements of high logical complexity. Some exciting work of H. Friedman points to the latter though to give a definitive answer seems to be premature.

16.1 REFERENCES

[1] Bishop, E., Foundations of Constructive Analysis. New York, McGraw-Hill, 1967.

[2] Feferman, S., In the Light of Logic. Oxford University Press. 340pp. (1998).

[3] Gerhardy, P., Kohlenbach, U., General logical metatheorems for functional analysis. To appear in: Trans. Amer. Math. Soc.

[4] Hilbert, D., Bernays, P., Grundlagen der Mathematik vol.I+II, 1934 and 1939. Springer Berlin.

[5] Kohlenbach, U., Effective moduli from ineffective uniqueness proofs. An unwinding of de La Vallée Poussin's proof for Chebycheff approximation. Ann. Pure Appl. Logic **64**, pp. 27–94 (1993).

[6] Kohlenbach, U., New effective moduli of uniqueness and uniform a–priori estimates for constants of strong unicity by logical analysis of known proofs in best approximation theory. Numer. Funct. Anal. and Optimiz. **14**, pp. 581–606 (1993).

[7] Kohlenbach, U., Analysing proofs in analysis. In: W. Hodges, M. Hyland, C. Steinhorn, J. Truss, editors, *Logic: from Foundations to Applications. European Logic Colloquium* (Keele, 1993), pp. 225–260, Oxford University Press (1996).

[8] Kohlenbach, U., Relative constructivity. J. Symbolic Logic **63**, pp. 1218-1238 (1998).

[9] Kohlenbach, U., On uniform weak König's lemma. Ann. Pure Appl. Logic **114**, pp. 103-116 (2002).

[10] Kohlenbach, U., Some logical metatheorems with applications in functional analysis. Trans. Amer. Math. Soc. **357**, no. 1, pp. 89-128 (2005)

[11] Kohlenbach, U., Effective uniform bounds from proofs in abstract functional analysis. To appear in: Cooper, B., Loewe, B., Sorbi, A. (eds.), 'CiE 2005 New Computational Paradigms: Changing Conceptions of What is Computable'. Springer Publisher.

[12] Kohlenbach, U., Applied Proof Theory: Proof Interpretations and their Use in Mathematics. Ca. 530pp, to appear in: Springer Monographs in Mathematics, 2008.

[13] Kohlenbach, U., Oliva, P., Proof mining in L_1-Approximation. Ann. Pure Appl. Logic **121**, pp. 1-38 (2003).

[14] Kohlenbach, U., Oliva, P., Proof mining: a systematic way of analysing proofs in mathematics. Proc. Steklov Inst. Math. **242**, pp. 136-164 (2003).

[15] Kreisel, G., On the interpretation of non-finitist proofs, part I. J. Symbolic Logic **16**, pp.241-267 (1951).

[16] Kreisel, G., Macintyre, A., Constructive logic versus algebraization I. In: Troelstra, A.S., van Dalen, D. (eds.), Proc. L.E.J. Brouwer Centenary Symposium (Noordwijkerhout 1981), North-Holland (Amsterdam), pp. 217-260 (1982).

[17] Simpson, S.G., Partial realizations of Hilbert's program. J. Symbolic Logic **53**, pp. 349-363 (1988).

[18] Simpson, S.G., Subsystems of Second Order Arithmetic. Perspectives in Mathematical Logic. Springer-Verlag. xiv+445 pp. 1999.

[19] Troelstra, A.S. (ed.) Metamathematical investigation of intuitionistic arithmetic and analysis. Springer Lecture Notes in Mathematics **344** (1973).

[20] Weyl, H., Das Kontinuum. Kritische Untersuchungen über die Grundlagen der Analysis. Veit, Leipzig 1918.

17

Penelope Maddy

Professor of Logic and Philosophy of Science
University of California at Irvine, USA

Why were you initially drawn to the foundations of mathematics and/or the philosophy of mathematics?

In high school, I attended an NSF summer school in mathematics that included a course on naïve set theory. When I learned that the number 1 could be *defined*, that $2 + 2 = 4$ could be *proved*, I was mesmerized. I read up on the construction of classical analysis in set theory, Russell's paradox and the need for axiomatization, and finally, the consistency and independence of the Continuum Hypothesis. Here was a simple question in the most fundamental of our mathematical theories, one that must have an answer (or so it seemed), and we didn't even know what an answer might look like! Hoping to learn more, I went to UC Berkeley to study set theory and foundations. There, in a set theory seminar, I was amazed and delighted by Scott's proof that the existence of a measurable cardinal implies that V isn't L. The conclusion was undoubtedly welcome, but I couldn't help asking, why *this* hypothesis? Questions like that eventually led me into philosophy.

What examples from your work (or the work of others) illustrate the use of mathematics for philosophy?

Formalization hasn't played a central role in my own work. David Malament's work in the foundations of physics seems to me an instructive example of how mathematics can be brought to bear on important philosophical questions.

17. Penelope Maddy

What is the proper role of philosophy of mathematics in relation to logic, foundations of mathematics, the traditional core areas of mathematics, and science?

To my mind, philosophy of mathematics is just one variety of metaphysics/epistemology: what is the nature of mathematical truth and how do we come to know it? Using the methods and resources of natural science seems to me the best way to approach such questions.

What do you consider the most neglected topics and/or contributions in late 20th century philosophy of mathematics?

What are the most important open problems in the philosophy of mathematics and what are the prospects for progress?

I think the most promising undertaking for now is careful case studies of how mathematical concepts are formed and how methodological decisions are made. Part of this project would be purely descriptive, but description as preparation for assessment: what makes this rather than that a good mathematical concept? What makes this methodological decision correct or incorrect? Study of particular cases might eventually lead to general principles of concept formation or methodology, but we shouldn't insist that such general theories are the only form success can take. It could be that the local analyses are the best answers to our questions, and in any case, they are almost certainly the best place to start.

18
Paolo Mancosu

Professor of Philosophy
and Chair, Group in Logic and Methodology
of Science
University of California at Berkeley, USA

Why were you initially drawn to the foundations of mathematics and/or the philosophy of mathematics?

Under the influence of my high school mathematics and physics teacher, I developed an interest in philosophy of science already in my last years in high school. I recall reading Popper and Bachelard in my senior year (1979). I thought philosophy of science would provide me with the right combination of humanistic and scientific knowledge and I decided to enroll in Philosophy at the Catholic University in Milan. Unfortunately, once I got there philosophy of science was no longer offered due to the departure of the faculty member who taught the subject. However, there was an excellent logician in the faculty who eventually became my (undergraduate) thesis advisor, Sergio Galvan. I recall my first course in logic as a real shock. It was an advanced course and except for me all the other students had already taken the course offered in the previous year, of which the one I was attending was the continuation. The topic was primitive recursive functions and their representability in Peano Arithmetic. Galvan was not one who would be satisfied with simply providing a sketch of the proofs. He went through all the details of the formalization. Particularly impressive was the painstaking development of the representability of Gödel's β-function. I managed to survive and in the following years I took more courses on topics such as the Hilbert-Ackermann Lemmas and on incompleteness and reflection principles in Peano Arithmetic. At the same time I had developed a strong interest in theoretical linguistics and for a while I was undecided as to which direction I should pursue. I opted for mathematical logic and I spent my last year in college (1983–84) writing my undergraduate thesis on "Models of Peano Arithmetic and Incompleteness Results". The thesis was divided into two parts. The first part was an overview of classical results on non-standard models of arithmetic. The second part gave an account of the recent 'indicator theory' developed by Paris and Kirby and of the Paris-Harrington theorem. In order to collect some of the recently published and unpublished work on the subject I also spent a month in 1983 at the University of Manchester where most of the work related to non-standard models of arithmetic was being carried out.

With such an interest in logic I could not fail developing also an interest in the philosophy of mathematics. While a student at the Catholic University, I also attended many courses at the State University of Milan in logic and in philosophy of science. Under the influence of Giulio Giorello, I became very interested in Lakatos and the history and philosophy of the infinitesimal calculus. The seeds of my later work on the seventeenth century (Mancosu 1996) were sown in his seminar on the philosophy of the infinite and the continuum. Of course, through other philosophy courses I was also able to see the immense importance of logic and the philosophy of mathematics for authors such as Husserl and Wittgenstein. However, I saw myself mainly as a logician and when I decided to apply to graduate school it was with the explicit intention of continuing in mathematical logic. Thanks to a Fulbright and an ITT grant I arrived at Stanford in 1984. I went through the superb sequence of offerings in logic (model theory, proof theory, recursion theory, set theory, constructive mathematics) offered at Stanford and I also took courses in mathematics and philosophy. Stanford was a very lively place in logic with, among others, Feferman, Barwise, Etchemendy and a constant stream of visitors. My interests were closer to Feferman's approach to logic and foundations. Under his guidance I wrote a dissertation in logic whose title was "Generalizing classical and effective model theory in theories of operations and classes" (Mancosu 1991b). This was a contribution to Feferman's program of 'explicit mathematics'.

Surprisingly, philosophy of mathematics was not offered at Stanford. In my four years at Stanford I was the only one to offer (together with Patricia Blanchette) an advanced seminar at the end of my studies there (1989). We covered material from the then recently published Tymoczko 1985. I also kept developing my interests in history and philosophy of mathematics, especially under the influence of the late Wilbur Knorr who encouraged my interests in the history of mathematics. By the end of my graduate studies I had already published an article on the debate on the infinitesimal calculus in the French Academy of Sciences in Paris (Mancosu 1989).

Looking then at my formative experiences retrospectively, I see how much of my published output was shaped by them. Through Galvan and Feferman I was able to develop my technical expertise in logic and foundations and with the help of Giorello and Knorr I pursued a historically inspired interest in philosophy of mathematics. My analytic interests in philosophy of mathematics are no less important but to a large extent I feel that I have been self-taught in that area.

What examples from your work (or the work of others) illustrate the use of mathematics for philosophy?

Let me begin with the most obvious interpretation of the question, i.e. as asking how the formal work I did in logic was a contribution to philosophy. When I worked in logic most of the drive I felt for the questions I was investigating came from the technical aspects of the questions and from the beauty I experienced in the subject matter. This was especially true of the work on non-standard models of arithmetic. The philosophical interest of the results, of which I was certainly aware, was a bonus but not the main motivation for the work. In the case of my Ph.D. dissertation, I happen to have chosen a topic which was part of Feferman's program of 'explicit mathematics' and thus whatever value my results have is as part of this program. Feferman's program of explicit mathematics can be briefly summarized as follows. Taking his start from Bishop's work on constructive analysis, Feferman noticed that Bishop-style constructive mathematics could be followed by classical mathematicians and read as a piece of classical mathematics, if they only chose to ignore the extra information (say, about the rate of convergence of a function) that a constructive mathematician feels obliged to provide. Feferman saw that one could develop a single formalism for capturing classical and Bishop-style constructive mathematics. The same strategy could be applied to areas of logic where constructive or effective analogues of classical notions had been developed. In particular, Feferman had developed an abstract theory of operations and classes for capturing what was common to classical model theory and model theory on admissible sets. In my dissertation I developed formal systems that could account for those generalizations of model theory studied by Feferman but that would also account for developments in recursive model theory. Philosophically speaking, the value of the program resides in isolating the 'effective' content of classical results in a unified framework, which can be understood by both the constructive and the classical mathematicians.

But there is another sense in which my work illustrates the importance of mathematics for philosophy. I think it is not an exaggeration to claim that all my work in history and philosophy of mathematics has been developed with the aim of studying key episodes showing the importance of philosophy for mathematics and of mathematics for philosophy. That mathematics has had a tremendous impact on philosophy is almost a platitude. My contribution has consisted in adding a number of detailed case studies of paradigmatic interest. Since my work has been mainly on the seventeenth (Mancosu 1996) and on the twentieth centuries (starting with Mancosu 1998), I will provide an example from both periods. The first example concerns the philosophical importance of Torricelli's discovery that a certain solid of infinite length has finite volume (Mancosu-Vailati 1991). This result was considered quite paradoxical at the time and forced many philosophers to revise or abandon their theories of infinity, their cosmologies, and their philosophies of mathematics. So here is a result in mathematics that deeply affected philosophy, if anything ever did. As an example from the twentieth century, I would like to quote the extended study that Tom Ryckman and I made of the relationship between Hermann Weyl and Oskar Becker (Mancosu-Ryckman 2002, 2005d). We investigated the major point of rupture between the claims of Husserlian phenomenology to be able to ground each cognitive act in intuition and the fact that both mathematics and physics had, by the 1920s, shown the limits of such ambitions. This realization led Becker and Weyl to abandon the Husserlian form of phenomenology presented in *Ideas* and to embrace either a new form of phenomenology (as in Becker's 'mantic phenomenology') or, as in Weyl's case, a 'symbolic construction of the world'. Once again, developments in mathematics and physics led to a major reshaping of philosophical theories. In conclusion, I think my work in history and philosophy of mathematics has added to our knowledge of the profound repercussions that developments in mathematics can have on philosophy.

What is the proper role of philosophy of mathematics in relation to logic, foundations of mathematics, the traditional core areas of mathematics, and science?

I see the role of philosophy of mathematics as being that of clarifying the philosophical presuppositions and consequences of the results obtained in these various areas. Moreover, I think that philosophy of mathematics should suggest problems that could be tackled rigorously with tools provided in the above-mentioned areas.

Let's begin with logic. The panorama of contemporary logics is very vast despite the predominance of first order logic. This predominance has been challenged from various corners and I see the goal of philosophy of mathematics here as that of articulating the rationale for preferring one logic over another. It may be objected that this should be the role of philosophy of logic but I would counter that 1) it is not easy to draw the boundary between logic and mathematics and 2) even if such boundary could be drawn, many issues in this area are motivated by philosophical concerns that relate to mathematics. Thus, to give the most obvious example, many claims in favor of second-order logic point out the inadequacy of first-order logic in capturing notions that are central to mathematics (see Shapiro 1991). Of course, the very claim I made about the vague boundary between mathematics and logic is a problem for the philosophy of mathematics (in connection with the philosophy of logic). In addition, the philosophy of mathematics can also have the role of suggesting the development of specific logical systems or metatheorems. Examples abound here but I would single out Hilbert's program as a program in philosophy of mathematics that led to numerous technical developments in logic itself (development of various formal systems, ε-calculus, decision procedures, completeness theorems, etc.)

Moving on to foundations of mathematics. It is of course not easy to make a clear cut distinction between philosophy of mathematics and foundations. But for the purpose of this discussion, I will take foundations to be what the discussion in the email list FOM is supposed to capture (whatever that may be, since there is no agreement among FOMers). According to this interpretation, foundations of mathematics is a particular area of mathematics that aims, using mathematical techniques, to clarify the foundations of the discipline. The variety of theorems pertaining to this area is immense and ranges from results about large cardinals, to subcubic graph numbers, to weak fragments of arithmetic. The role of the philosophy of mathematics is twofold here. First, it seems to me that the philosopher of mathematics should try to articulate the philosophical importance of the plethora of results that have been achieved in foundations. For instance, there has been extensive proof-theoretic work on the reducibility of non-predicative systems to predicative ones. Few philosophers of mathematics are conversant with such results and the task of articulating their philosophical relevance has barely begun. Secondly, the philosophy of mathematics should always try to formulate philosophical theses in ways that can be precisely analyzed and possibly tackled mathematically. While I do not believe that all interesting problems in philosophy of mathematics can be turned into precise problems of this sort (more about this below), it is certainly a boon when such precision can be reached. Thus, I found the technical work that Boolos and others carried out in connection to 'Frege's theorem' to be extremely satisfactory and rewarding for a proper understanding of Frege's program in philosophy of mathematics and for discussions on neologicism. Another paradigmatic example of a successful formal analysis of an idea originating in the philosophy of mathematics is the characterization of predicative provability given by Feferman and Schütte in the 1960s as providing a precise articulation of informal comments on predicativity made by Poincaré and others.

Traditionally, the philosophy of mathematics had been developed as a collection of theses concerning several areas of mathematics. In the seventeenth century, for instance, different accounts were given for arithmetic, algebra and geometry. In the early part of the twentieth century the unifying power of set theory led to philosophy of mathematics being virtually reduced to the philosophy of set theory. A major problem here concerns the nature of this reduction and whether all of mathematics can be reduced (in whatever sense) to set theory. Challenges to the last claim have come, for instance, from category theory. But apart from the discontent expressed by the category-theory community there has been little protest when it comes to the philosophical propriety of addressing issues in philosophy of mathematics by translating them as issues in philosophy of set theory. I can explain the source of my worry here by recalling another seventeenth century example. Newton and Leibniz were of course master practitioners of analytic geometry. In many ways, what Descartes had achieved was showing that all traditional geometry could be captured analytically. However, both Leibniz and Newton thought that the reduction also amounted to a loss and they resisted the temptation to treat problems about geometrical constructions in purely analytic terms. While aware of the advantages of the reduction of classical geometry to analytic geometry they also realized that each area had its own methodological autonomy and was the source of independent technical and philosophical problems. This has not been, by and large, what has happened in contemporary philosophy of mathematics. Philosophers rarely provide specific methodological analyses of core areas of mathematics such as complex analysis, real analysis, combinatorial topology, etc. The dominant paradigm has been that of accepting the reduction of all such areas to set theory and to pose all the philosophical problems in those terms. But there is room for questioning whether this has not dramatically limited the ability of philosophy of mathematics to account for mathematical practice. In particular, the reduction in principle to set theory does not imply that the disciplines in question are methodologically on a par nor that the problems they give rise to are identical. The situation has only recently begun to change with a new generation of philosophers of mathematics moving away from this reductionist paradigm.

Finally, what should the relation between philosophy of mathematics and the sciences be? The major goal for philosophy of mathematics in this area is that of accounting for the applicability of mathematics to the sciences. A related important problem concerns whether, or how much, mathematics is indispensable for science (and this independently of whether one accepts the Quine-Putnam indispensability argument). Much work needs to be done in this area as we lack detailed case studies. To give just one example, consider the role mathematics plays in scientific explanations. Sometimes the proper explanation of a scientific fact seems to rely on the geometry (or the arithmetic) of the situation. In such cases many people talk of a mathematical explanation of the scientific fact. We have no satisfactory analysis of what is going on in such cases of mathematical explanation. The problem is not only relevant to how mathematics 'hooks on' to reality (the applicability problem) but also to recent versions of the indispensability argument which give up the original holism contained in the Quine-Putnam version of the argument. In addition to the problems I just mentioned, there is also a vast continent to be explored in the analysis of the way in which theoretical physics is leading to new results in mathematics (see Jaffe 2004). This has captured the attention of several French philosophers of mathematics (such as Jean Petitot) who seem to think that these new developments justify a renewed form of Kantianism.

What do you consider the most neglected topics and/or contributions in late 20th century philosophy of mathematics?

I think that philosophy of mathematics has to a great extent been hijacked by metaphysics and by the particular position of the problem that finds its clearest expression in Benacerraf's well known articles. In recent decades, this has had as a consequence an extremely narrow view of mathematical epistemology within mainstream philosophy of mathematics, a view partly due to the over-emphasis on ontological questions. For the most part, current epistemology of mathematics has not addressed at all matters relating to fruitfulness, understanding, explanation and other aspects of mathematical epistemology. Only recently a new generation of philosophers of mathematics has began to address such issues (see Mancosu 2005a, 2007). Also related to epistemology is the need for developing an epistemology of mathematics that will fit well with what we know about human vision and cognition. Despite endless appeals to 'naturalism' there is a frightful scarcity of good philosophical work providing a 'naturalistic' account of mathematical knowledge. The work I found most promising and inspiring in this direction is that of Giaquinto on perception and visualization in mathematics (see Giaquinto 2007); he starts from vision science and ends up providing an account of mathematical knowledge.

What are the most important open problems in the philosophy of mathematics and what are the prospects for progress?

As a consequence of the issues discussed in section four, I advocate a philosophy of mathematics that is closer to mathematical practice than is currently the case. As I see it, philosophy of mathematics has paid very little attention to mathematical practice, insisting on theorizing while accommodating relatively few facts about actual mathematics. It should be the other way around, with constrained and possibly tentative philosophical conclusions resting on a deep and extensive base of knowledge about mathematics. In addition, our theorizing should not ignore the essentially diachronic nature of the subject. This change of perspective gives rise to a vast field of areas and a large number of open problems.

In order to be more specific about the sort of topics one can fruitfully investigate, let me mention the rationale for a book on "The Philosophy of Mathematical Practice" which I am editing for Oxford University Press (Mancosu 2007). The book is the first attempt to give a coherent and unified, although not exhaustive, presentation of this new direction of work in philosophy of mathematics (see also Mancosu 2005a). This new approach in philosophy of mathematics requires extensive attention to mathematical practice. Of course, this is not to say that previous developments in philosophy of mathematics were completely removed from such concerns. Some attention to mathematical practice is found in the classical foundationalist programs (logicism, intuitionism, Hilbert's program) and to a certain extent also in the tradition of analytic philosophy of mathematics.

In addition, mathematical practice has also been a concern of the tradition originating from Lakatos 1976 which in the Anglo-American literature gave rise to such contributions as Kitcher 1983, Aspray and Kitcher 1988, Tymoczko 1985, Gillies 1992, Grosholz, Breger 2000, and now Ferreiros, Gray 2006. What the contributors to these volumes called for was an analysis of mathematics that could account for its historical nature and development. The questions that interested them were, among others: How does mathematics grow? How are informal arguments related to formal arguments? How does the heuristics of mathematics work and is there really a sharp boundary between method of discovery and method of justification? However, even this 'maverick' tradition (to use a term from the Aspray-Kitcher introduction to Aspray-Kitcher 1988) was limited in its attention to mathematical practice (an exception here is the recent book Corfield 2003).

Finally, even within the more traditional analytical approaches to philosophy of mathematics the discussion has in some cases led naturally to developments that squarely address issues of methodology of mathematics. One of the best known cases is that of Penelope Maddy (see for instance, among her many contributions, Maddy 1997) who has investigated the considerations by which mathematicians accept new axioms etc. Her case studies come from set theory but her approach can be generalized to other areas and indeed a recent issue of *Logique et Analyse* (van Kerkhove and van Bengedem 2002) contains a significant number of contributions along these lines.

In what sense then does the more recent attention to mathematical practice, which the book I am editing intends to present, mark a change here? At least in two senses. The first is the claim that there are important *novel* characteristics of contemporary (twentieth century) mathematics that are just as worthy of philosophical attention as at the time of the foundational debates the distinction between constructive vs. non-constructive, etc. Examples recently discussed in the literature include category theory, cohomology theory and knot theory. Secondly, traditional (foundationalist) philosophy of mathematics has often been carried out within the rigid boundaries of mathematical logic. Philosophical issues that escaped formal treatment – such as visualization, explanation or other cognitive issues related to mathematical understanding – were as a consequence relegated to the dustbin of the 'subjective' and thus to the area of the philosophically uninteresting. In particular, my own work has centered on 'mathematical explanation'.

If mathematicians only cared about the truth of certain results it would be hard to understand why, after discovering a certain mathematical truth, they often go on to prove the result in several different ways. This happens because different proofs or different presentations of entire mathematical areas (complex analysis etc.) have different epistemic virtues. 'Explanation' is among the most important epistemic virtues that mathematicians seek. Very often the proof of a mathematical result convinces us *that* the result is true but does not tell us *why* it is true. Alternative proofs or alternative formulations of entire theories are often given with this explanatory aim in mind. The topic of 'mathematical explanation' (Mancosu 2001, 2005b) is also intimately related to issues in philosophy of science via the notion of 'scientific explanation'. I do not see explanation as an insular subject but rather as only a component part of a general account of mathematical understanding.

Finally, let me return to the general structure of the book I am editing with the list of the eight topics I have selected and the names of the scholars writing on them. This list will give the reader an idea of the broad spectrum of the recent philosophical reflection on different aspects of mathematical practice: Diagrammatic reasoning (Ken Manders); Visualization (Marcus Giaquinto); Mathematical Explanation (Paolo Mancosu); Purity of Methods (Mic Detlefsen); Mathematical Concepts (Jamie Tappenden); Philosophical relevance of category theory (Colin McLarty); Philosophical aspects of computer science in mathematics (Jeremy

Avigad); Philosophical impact of recent developments in mathematical physics (Alasdair Urquhart). I strongly believe that the book gives voice to a promising direction of research that will yield rich dividends.

Caveat lector: although my emphasis has been on a new direction in the philosophy of mathematical practice, from this it should not be inferred that I underestimate the importance of the work that remains to be done in foundations or in analytic philosophy of mathematics. Rather, it is meant to emphasize a direction of work that has so far been unjustly ignored.

Bibliography

Aspray, W., and P. Kitcher, eds., 1988, *History and Philosophy of Modern Mathematics*, University of Minnesota Press.

Corfield, D., 2003, *Towards a Philosophy of Real Mathematics*, Cambridge University Press.

Gillies, D., ed., 1992, *Revolutions in Mathematics*, Oxford University Press.

Ferreiros, J., and J.J. Gray, eds., 2006, *The Architecture of Modern Mathematics*, Oxford University Press.

Giaquinto, M., 2007, *Visual Thinking in Mathematics: an epistemological study*, forthcoming for Oxford University Press.

Grosholz, E., and H. Breger, 2000, *The Growth of Mathematical Knowledge*, Kluwer.

Jaffe, A., 2004, Interactions between mathematics and theoretical physics, in T. H. Kjeldsen et al., *New Trends in the History and Philosophy of Mathematics*, University Press of Southern Denmark, Odense, 2004, pp. 87–103.

van Kerkhove, B., and J.P. van Bengedem, eds., 2002, *Perspectives on Mathematical Practices*, Special issue of *Logique et Analyse*, 179–180, 2002 [but published in 2004]

P. Kitcher, 1983, *The Nature of Mathematical Knowledge*, Oxford University Press.

Lakatos, I., 1976, *Proofs and Refutations*, Cambridge University Press.

Maddy, P., 1997, *Naturalism in Mathematics*, Oxford University Press.

Mancosu, P., 1989, The metaphysics of the calculus: a foundational debate in the Paris Academy of Sciences, 1700–1706, in *Historia Mathematica*, 16, pp. 224–248.

Mancosu, P., 1991a, (with E. Vailati), Torricelli's infinitely long solid and its philosophical reception in the XVIIth century, *ISIS*, 82, pp. 50–70.

Mancosu, P., 1991b, Generalizing classical and effective model theory in theories of operations and classes, *Annals of Pure and Applied Logic*, 52, 3, pp. 249–308.

Mancosu, P., 1996, *Philosophy of Mathematics and Mathematical Practice in the Seventeenth Century*, Oxford University Press.

Mancosu, P., ed., 1998, *From Brouwer to Hilbert. The Debate on the Foundations of Mathematics in the 1920s*, Oxford University Press.

Mancosu, P., 2001, Mathematical Explanation: problems and prospects, *Topoi*, 20, pp. 97–117.

Mancosu, P., 2002, (with T. Ryckman), Mathematics and Phenomenology. The correspondence between Oskar Becker and Hermann Weyl, *Philosophia Mathematica*, 10, pp. 130–202.

Mancosu, P., 2005a, co-edited with K. Jørgensen and S. Pedersen, *Visualization, Explanation and Reasoning Styles in Mathematics*, Springer, pp.x+300.

Mancosu, P., 2005b, (with J. Hafner), The varieties of mathematical explanation, in P. Mancosu, K. Jørgensen and S. Pedersen eds., *Visualization, Explanation and Reasoning Styles in Mathematics*, Springer, pp. 215–250.

Mancosu, P., 2005c, (with T. Ryckman), Geometry, Physics and Phenomenology: the correspondence between O. Becker and H. Weyl, in *Die Philosophie und die Mathematik: Oskar Becker in der mathematischen Grundlagendiskussion*, ed. Volker Peckhaus, Wilhelm Fink Verlag, München, 2005, pp. 153–228.

Mancosu, P., 2007, ed., *The Philosophy of Mathematical Practice*, forthcoming for Oxford University Press.

Shapiro, S., 1991, *Foundations without Foundationalism*, Oxford University Press.

Tymoczko, T., ed., 1985 (2nd edition, 1998), *New Directions in the Philosophy of Mathematics*, Birkhäuser.

19

Charles Parsons

Edgar Pierce Professor of Philosophy, Emeritus
Harvard University, USA

Why were you initially drawn to the foundations of mathematics and/or the philosophy of mathematics?

Although I had some interest in philosophy as an adolescent, I really began with mathematics. I began to take a serious interest toward the end of high school and in the course of my freshman year of college decided on a mathematics major. However, I always doubted that I would become a professional mathematician, and one motive was the thought that mathematics might be a good foundation whatever field I went on in. My undergraduate work did not at all emphasize logic or foundations of mathematics, although early on I studied W.V. Quine's *Methods of Logic* on my own, and I took his more advanced logic course as a junior. (I have written about this in "W. V. Quine: A student's-eye view," *Harvard Review of Philosophy* 10 (2002), 6-10.) In the philosophy that I took, I was exposed to the importance logic had for analytic philosophy but not to the philosophy of mathematics *per se*. Toward the end of my undergraduate career I decided to study philosophy but was not settled on any particular interest or direction.

It was events of the next 2 1/2 years that drew me into foundations of mathematics. During the year I spent at the University of Cambridge right after college, I attended lectures on logic by James Thomson and S.W.P. Steen, the latter my first exposure to a proof-theoretic point of view. (Quine had taught the incompleteness theorem but did not convey its origin in Hilbert's program or other related results in that tradition.) Margaret Masterman's lectures and conversation urged on me the importance of Brouwer, though her own motivations came from the philosophy of language. I decided to spend my last term at Cambridge studying

Brouwer's writings, particularly papers of the 1920s presenting central ideas of intuitionistic mathematics from his mature perspective. When I returned to Harvard in 1955 as a Ph. D. student, I already had an idea of the main tendencies in the foundations of mathematics. During my first year I took a seminar and a reading course with Hao Wang; the seminar had some philosophical content but was mainly logical, and the reading course was on Kleene's *Introduction to Metamathematics*. I had also by that time come to know of Kreisel's early work but had not studied it.

I was by then inclined to turn to the foundations of mathematics for a dissertation. The experiences just described had made me aware that there were alternatives to logicism and logical positivist views, which dominated Anglo-American philosophical discussion of mathematics at the time. I had also been studying Kant and thought of a philosophical dissertation in some way relating Kantian ideas and the different tendencies in foundations. I would probably have had a hard time finding an adviser for such a project, particularly since Wang left Harvard in 1956. When I sought out Burton Dreben, who joined the faculty that fall, he urged me to do a thesis in mathematical logic. I think I was easy to persuade; clearly work in logic would be a good foundation for whatever I might aspire to do in philosophy of mathematics.

I did not want to be enlisted in Dreben's own program on the decision problem for first-order logic. It was too remote from my philosophical interests and did not excite me as mathematics. Dreben was sympathetic to my interest in proof theory, and that is the direction I took, taking as my point of departure Hilbert and Bernays' *Grundlagen der Mathematik* and the early papers of Kreisel. But the framework for the central parts of my dissertation came from papers of Kurt Schütte that I had first heard of through Steen's lectures in 1954–55.

The above does more to say *how* I was drawn into foundations of mathematics than why. That's a difficult question. One thing is clearly that it offered a way to combine the two intellectual subjects that most engaged me, mathematics and philosophy, especially since I thought I did not have the temperament of a mathematician or an outstanding talent for mathematics. But I don't think it was an *ad hoc* combination. A lot of philosophical discussion after the rise of mathematical logic centered on the question of the limits of logic. Had it really been proved, as was often claimed, that logicist and set-theoretic constructions made Kant's pure intuition, or other ideas deriving from it, otiose? Some of my early philosophical writing questioned that, and proof theory, my central concern as a logician up until 1973, arose from a tradition that also questioned this conclusion.

What examples from your work (or the work of others) illustrate the use of mathematics for philosophy?

Almost all major rationalist philosophers illustrate a mining of mathematical knowledge and training in philosophy. But I have been more influenced by those who did actual mathematical work (including logical) with the aim of clarifying philosophical questions. The examples from the foundations of mathematics are too well-known: Frege, Whitehead and Russell, Brouwer, Hilbert, Ramsey, Carnap, Quine. I hope that some of what has been done by persons close to my own generation and later has the same character, if not the same classic status.

A rather complex case is a figure on whom I have done scholarly work and of whose posthumous works I was a co-editor: Kurt Gödel. I think he thought of his great early theorems primarily as mathematical results. Although in retrospective comments he stressed the heuristic role of his philosophical attitude, this would be a case of the use of philosophy for mathematics rather than the other way around. However, in his later lectures and writings on the philosophy of mathematics he often made use of his own theorems in arguments.

Also very well known, but off the subject of philosophy of mathematics, is the use in philosophy in general of logical tools such as modal logic and its various developments.

Since my work in philosophy of mathematics consists largely of reflection *about* mathematics and logic, in what sense is it a use of mathematics for philosophy? Does the philosopher of perception use perception for his philosophy? In a way he does, because he probably will mine his own experience as a perceiver and not just the language and science of perception, the writings of other philosophers, and his own ratiocination. I have appealed to mathematical results (especially in logic) at various points, for example to distinguish different senses or levels of predicativity.

What is the proper role of philosophy of mathematics in relation to logic, foundations of mathematics, the traditional core areas of mathematics, and science?

That's a very large question. I'll begin with the easiest part. "Foundations of mathematics" is a term that has been used in different ways, so that it might include philosophical work (even most philosophy of mathematics), but it very often includes a lot of work in mathematical logic (and maybe some other mathematical work), and some use it exclusively for mathematical programs. I would see philosophical reflection as part of that enterprise, even where a mathematical program is involved.

As regards logic, there is of course philosophy of logic as well as philosophy of mathematics, while logic itself is largely part of mathematics, though some have described it as applied. There isn't a sharp line between logic and mathematics, so that there shouldn't be a sharp line between philosophy of logic and of mathematics either. Even elementary logic is in a certain way entangled with mathematics, so that considerations about mathematics, possibly not immediately about logic, can be relevant to philosophical reflection about logic.

About the "traditional core areas of mathematics," I am handicapped because my own work has been shaped by mathematical logic and the earlier axiomatizations of mathematics that logic then worked with. There are other kinds of work that take informal mathematics much more at face value or engage the earlier history of mathematics. I have not done enough such work to say what the possibilities are or what the proper role of philosophy should be. I don't see philosophy of mathematics as part of a "first philosophy" that can legislate for mathematical practice. But few of us do.

About science, the application of mathematics in science is a natural subject for philosophers to be concerned with; the questions presumably belong both to the philosophy of mathematics and the philosophy of science.

What do you consider the most neglected topics and/or contributions in late 20th century philosophy of mathematics?

I will comment on neglected topics (i.e. inquiries little undertaken) rather than on neglected contributions (i.e. philosophical work to which little attention has been paid).

Philosophically informed studies of the history of mathematics were certainly neglected in the sense of being little done during at least the middle part of the twentieth century, until well after I myself started out. This has gradually changed, especially as regards the nineteenth-century revolution in mathematics and its aftermath in foundational studies. But there is probably still a lot of relatively uncharted territory elsewhere, in particular in twentieth-century mathematics outside what we normally think of as foundations.

Another line of inquiry that was neglected was reflection on the psychological studies of the development of numerical and other mathematically relevant competence in young children. Probably this would illuminate only the most elementary mathematical knowledge, but one never knows.

To turn to something closer to home, mathematical logic, a number of philosophers are well educated in proof theory, or set theory, or computability theory. It seems to me that fewer are well educated in model theory. Since my knowledge of that subject is deficient, I can't be sure how much difference such education would make to philosophical work. But it seems likely that it would make some.

What are the most important open problems in the philosophy of mathematics and what are the prospects for progress?

I believe that doing justice to mathematical knowledge requires some concession to a rationalistic point of view. In my own work I came to this conclusion rather late, and a robust rationalist would find my own concession minimal and no doubt inadequate. However that may be, the search for the right balance between rationalistic and other considerations will go on.

Furthermore, rational knowledge seems to me not well understood even after a rationalistic tradition going back to Plato. That's a subject that deserves more exploration.

About the prospects for progress, the history of the debate between rationalists and their opponents is not encouraging. Possibly there are limits to what can be done to advance from description to explanation. But we are unlikely to discover them without trying to overcome them.

The higher infinite and the frontiers of set theory more generally pose a standing epistemological challenge. Although it's difficult for philosophers to keep up with what the set theorists do, the philosophical rewards might be great.

Although I won't claim that it is one of the most important problems, here is an unpaid debt of my own work and that of some others on a structuralist view of mathematical objects. Suppose we have a mathematical theory that intuitively describes a structure such as the natural numbers, Euclidean space, or some universe of sets, which is not constructed from some more basic structure. Structuralists have said that the objects of the structure exist if the theory is *coherent*. But a lot more needs to be said about what that means than is said in my own writings or in the literature that I know.

20
Michael D. Resnik

University Distinguished Professor, Emeritus
Department of Philosophy
University of North Carolina at Chapel Hill, USA

Why were you initially drawn to the foundations of mathematics and/or the philosophy of mathematics?

After reviewing my answer below, I see that my account is more an autobiographical account of *how* I was drawn to work in philosophy of mathematics and how my views developed instead of an account of *why* I was motivated to do this work. Neither my "How?" answer below nor my "Why?" answer (not elaborated upon here) is particularly deep. The short of both is that through a series of events I was exposed to these subjects, and found working in them exciting.

Here is a longer answer to the "How"-question.

I took my first logic course from Alan Anderson in the spring of 1957 while in my first year as a Yale undergraduate. We used Copi's *Introduction*, including the little bit it contains on symbolic logic. I liked doing simple problems in logic so much that I took a year long course from Frederic Fitch the next year, and then continued taking every logic course I could from Anderson and Fitch. In order to do this I majored in *Mathematics and Philosophy*, which in turn required me to take Kakutani's course in analysis. Every year there was one student in the course, the star, who went on to do important work in mathematics. David Krantz and Richard Beals starred in Kakutani's course when they took it. I was no star, to say the least, and taking that course made it clear to me that I was not cut out for a career in mathematics. But I still loved studying and "doing" logic, so considered doing graduate work in philosophy.

A fellow student told me to read Quine's *From a Logical Point of View* and Ayer's *Language, Truth and Logic*, and when I did

I knew that pursuing philosophy was the right thing for me. I was accepted into the Ph.D. program at Harvard, and started my studies there in the fall of 1960, taking a set theory course from Quine. I passed the preliminary exams at the end of my first year, and started looking for a dissertation topic. Someone told me to speak to Burton Dreben, whom I had not yet met. He showed me his work on solvable sub-cases of the decision problem for the predicate calculus and invited me to work on some open problems with him. Once again my lack of mathematical talent stood in the way, but along the way I learned a lot of meta-logic from Dreben, Quine and Hao Wang.

Dreben's course in the spring of 1962 on Frege showed me how to combine my interest in logic with my ability to understand complicated technical matters and to write clear explanations of them. I did not realize until later that this was one of my most important talents.

With Dreben's encouragement I started to work on a dissertation expounding Frege's methodology, its philosophical presuppositions, and how these led to certain features of his symbolic logic. A year and one half later, I completed this dissertation under the joint direction of Charles Parsons and Dagfinn Føllesdal. When I left my short stay at Harvard at the end of 1963 I really did not know much philosophy or much about how to do research in the subject. But I had been much attracted to Quine's work in the course of writing my dissertation and I continued to read his writings. I also wrote several papers on Frege, and began to think about philosophical issues in mathematical logic and the foundations of mathematics. I think at that time I was strongly attracted to the constructivist spirit that was embodied in the proof theory I had been reading.

In my readings I came upon the claim made by some logicians that the Löwenheim-Skolem Theorem showed that every set was countable. After examining the various proofs of the Theorem, I concluded that the claim was false. As I noted, even the proof using the axiom of choice to cut down the sets of a model until only countable ones remain does not establish that the uncountable sets of the original model are identical to the countable ones in the cut down model. This was the thrust of my paper "On Skolem's Paradox". The paper drew a response from my friend William Thomas, who referred to my "creeping Platonism". I think it was then that I realized that a deeper view of the Theorem required one to take a philosophical stand on the nature of mathematical language and its subject matter (if any). It was then that my serious thinking about non-formal matters in philosophy began.

Writing *Frege and the Philosophy of Mathematics* in the late 1970's was a wonderful way for me to explore the various traditional approaches to the philosophy of mathematics and to see their advantages and drawbacks. I came away realizing that the most important lesson of the work in the philosophy and foundations of the first half of the Twentieth Century was that infinity does not come cheap. Axioms of infinity are essential to any viable foundational system; and knowledge of infinite structures, or even our knowledge that there are infinitely many mathematical objects is problematic.

Paul Benacerraf's presentation to the American Philosophical Association of "Mathematical Truth" and Oswaldo Chateaubriande's comments on it led me to the idea that mathematical knowledge could be based upon knowledge of patterns. At first I probably thought that we could learn about patterns from experience and imagination, but that seemed to fail to account for our knowledge of infinite patterns. Later I realized that Quine's holism could provide for our knowledge of infinite patterns, and I crafted a holist epistemology of mathematics.

What examples from your work (or the work of others) illustrate the use of mathematics for philosophy?

First off, I would distinguish between *doing* mathematics to advance a philosophical position, *citing* mathematics to buttress a philosophical thesis or argument, and *reflecting* philosophically upon mathematics. Frege and Russell, and Hilbert *did* mathematics in the service of philosophical theses. Frege even said at the end of his *Foundations of Arithmetic* that to demonstrate that arithmetic is a branch of logic it is necessary to carry out a formal deduction of the former from the latter. And Hilbert developed proof theory to support a philosophical claim, namely, that from the finitary standpoint one can justifiably use infinitary mathematics. Turning to more recent work, I would also count Hartry Field's *Science without Numbers*, Geoffrey Hellman's *Mathematics without Numbers*, and part of Charles Chihara's *Constructibility and Existence* as the results of philosophers doing mathematics to advance their philosophical positions. (Of course, I am assuming that in constructing a foundational system one is doing mathematics—something that various mathematicians would controvert!) On the other hand, one might *cite* Russell's Paradox and Godel's Incompleteness Theorems as bits of mathematics that tend to refute the philosophical claims of Frege and Hilbert, respectively. Similarly, one might indicate some mathematical feature of quantum mechanics as tending to undercut Field's program. Or one might *cite* the indispensability of assumptions of infinity for reconstructions of classical mathematics in criticizing an ontology or epistemology for mathematics. Thus it is usual to point out that Nelson Goodman's nominalist ontology cannot provide for classical analysis. Again one might cite the same mathematics in setting out an ontology or epistemology for mathematics. E.g., recognizing the need for some kind of infinite led Chihara and Hellman to add modalities to the nominalist ontologies. Finally, in ones capacity as a philosopher one might *reflect* upon the issues raised by the Church-Turing Theorem (and the Church-Turing Thesis), or Godel's Incompleteness Theorems, or Skolem's Paradox. As I noted in answering question #1, my essays on Skolem's Paradox and Godel's Incompleteness Theorems have been of this kind.

What is the proper role of philosophy of mathematics in relation to logic, foundations of mathematics, the traditional core areas of mathematics, and science?

This is a normative question, calling for a value judgment. With this in mind, let us note that there are philosophical controversies over whether philosophy has any role to play with respect to mathematics or science. Penelope Maddy, for example, has argued that despite the philosophical debates between mathematicians over non-constructive proofs and definitions, the axiom of choice, etc. mathematics proceeds and *should* proceed without the guidance of philosophy. On her view, philosophy cannot legitimately critique mathematics, empirical science or their methodologies. I think there is a bit of truth in what she says: Philosophy has no access to a priori norms through which to stand back and assess mathematics and science, for there are no such norms. On the other hand, the practice of philosophy blends with the practice of science and mathematics with the result that it may be unclear whether one is functioning as a philosopher or as a mathematical or empirical scientist. When Einstein objected to quantum mechanics on the ground that "God does not throw dice", was he speaking as a philosopher or as a physicist? Were Gödel's remarks about the meaningfulness of higher set theory made from a mathematical or a philosophical point of view? Maddy thinks of Quine as a meddling philosopher when he urged that the scientifically unnecessary speculative reaches of the set theoretic hierarchy be deactivated by adding the axiom of constructability, $V = L$, to axiomatic set theory. According to Maddy, most set theorists disagree with such a restrictive approach to set theory. But whether Quine was right or wrong, it is not clear that here we have a philosopher disagreeing with mathematicians. Maybe we just have one set theorist – remember Quine worked extensively in this area – who is out of step with the majority of set theorists. The point I am trying to make, however, is that without a clear distinction between philosophy (or philosophers), on the one hand, and mathematics and science, (or mathematicians and scientists), on the other hand, the claim that philosophy has no role to play in science and mathematics does not make sense.

Perhaps, it's not the professional credentials of the person that matter but rather the sort of reasons they give. But what makes a reason count as philosophical as opposed to scientific? If we look at the meta-scientific debates that have been classed as philosophical, they are all normative. So we might count a normative reason as a philosophical one. But then all discussions of the direction a scientific or mathematical field should take would count as philosophical. Indeed, even the reason that set theorists reject $V = L$ (it is too restrictive) would count as philosophical. The upshot seems to be that the question of whether *philosophy* has any role to play in mathematics or science is really the question of whether *professional philosophers*, who are not themselves professional mathematicians or scientists, have any role to play in these fields.

I think the answer is that they do. With sufficient training in mathematics and science, philosophers may be well placed to evaluate results, to criticize failures of clarity or rigor, and to prod and reform science and mathematics. It's not that their pronouncement should trump all others, but only that they should not be dismissed out of hand.

Philosophers also may be able to correct overblown renditions or misinterpretations of mathematical and scientific results. Consider the claims that have been made of the import for the human mind of the Unsolvability of the Halting Problem or Godel Incompleteness Theorems and the philosophical "therapy" imparted in response to these claims.

Philosophers can prompt mathematical research—the programs

of the early 20th Century illustrate this. Philosophers can motivate the development of new "logics" and new foundational systems. Philosophers often provide new ideas in rough form that mathematicians can make precise and develop deductively.

What do you consider the most neglected topics and/or contributions in late 20th century philosophy of mathematics?

One of the central questions of 20th Century Philosophy of Mathematics has been the relation of logic to mathematics. Frege, Russell, Brouwer, Heyting, Hilbert, Carnap, Quine and Wittgenstein have all written on the topic. Each worried about the status of logic—what its scope and limits are, and whether it had any authority over or role to play in mathematics. Late 20th Century philosophy of mathematics seems to have lost sight of the issues these great figures raised. For most work in recent philosophy of mathematics has simply assumed that if some aspect of mathematics can be shown to be a matter of logic then important progress has been made. Of course, this attitude underlay Frege's version of logicism, but more recently we find it in the work of Hartry Field, Geoffrey Hellman, and the neo-logicism of Crispin Wright and Bob Hale. Each of these philosophers has assumed that the nature of logic is clear and that it is epistemically superior to mathematics.

However, serious issues about the scope and limits of logic remain open. Does it include second-order logic? Is there but one logic? Are the rules of logic open to revision or expansion? How does the interplay between our normative judgments – our intuitions of correct inference – and our descriptive metaphysical judgments – our intuitions of what is possible or necessary – work in the epistemology of logic?

Suppose that we take these serious issues as settled. It is still not at all clear that mathematicians proceed by logically deducing new theorems from axioms or previously accepted results. Moreover, even if the theorems of mathematics can be reconstructed as the logical consequences of accepted axioms, it is not clear what this shows about the subject matter of mathematics or mathematical knowledge. After all, much of mathematics can be recaptured within a geometrical framework. Does this show that such mathematical knowledge is geometrical knowledge?

Finally, there are questions concerning specific attempts to reduce mathematics to some kind of logic basis. For example, both

Field and Hellman argue that we cannot know mathematical truth as construed by mathematical realists. Yet they believe that we do know (or at least can know) that ZF is logically consistent. But they provide little information on how we can know this.

What are the most important open problems in the philosophy of mathematics and what are the prospects for progress?

I think that the most important open problems in the philosophy of mathematics are the perennial ones – the nature of mathematics – what it is about – and the nature of mathematical knowledge. Thanks to the extensive work in foundations, we know much more about the demands on a philosophy of mathematics than our predecessors did. It's clear to us that the need for axioms of infinity and reducibility signify the failure of Russell's program. We know why Locke's way of ideas leaves out so much and why Kant's and Mill's approaches will not work, and why Frege's approach can go only so far. Our knowing more about why historical approaches have not worked is knowledge our predecessors lacked. In this we have a mark of progress in the philosophy of mathematics. Of course, I have offered solutions to these perennial problems myself in arguing that mathematics is a science of patterns, and many of my contemporaries have offered their own solutions. However, no "solution" has emerged unscathed by philosophical criticism, and most have only the support of their authors. So while I expect that new and ingenious "solutions" will be proposed, I don't expect these problems to be permanently resolved. This is the nature of philosophy.

21
Stewart Shapiro

O'Donnell Professor of Philosophy
Ohio State University, USA
Professorial Fellow
Arché Research Centre, University of St. Andrews, UK

Why were you initially drawn to the foundations of mathematics and/or the philosophy of mathematics?

As an early teenager, I showed aptitude for mathematics. My inner city middle school had a wonderful teacher, named Sam Traficant, who encouraged me to pursue my interests. I spent many hours in Mr. Traficant's classroom, after school, being introduced to such topics as alternate notations, calculus, and continued fractions. I devoured popularizations of advanced mathematics, and was hooked. In retrospect, I was taken by the (at least apparent) certainty of it all, with how it is possible to provide a sustained argument establishing a conclusion definitively, beyond any rational doubt. In short, I became fixated with proof.

My fascination was enhanced, and focused, when I attended a summer program for high school students held at Ohio State University. The director, Arnold Ross, gave a course in number theory, where we were introduced to topics like quadratic reciprocity. Professor Ross displayed a remarkable, and unique teaching style. Every lecture was accompanied with a list of problems, some of which were open, at least as far as the program was concerned. Once the list was distributed, the students, college-age counselors, and junior faculty would work on the problems together, sometimes contributing and combining partial results. I was shown then, by example, how learning can be done cooperatively, better than in the more usual competitive environments. One main focus of the program was on what Dr. Ross called "tight proofs", which are deductive discourses that leave no gaps, or at least few gaps, to be filled. This, I take it, was the main impetus to the longstanding thesis that logical consequence is matter of form. Another emphasis of the program was on the foundations of mathematics, which consisted of working toward general theorems from basic axioms. Dr. Ross's slogan, "think deeply of simple things", is a fine guide for the philosopher as well as the mathematician.

In my second and third summer in the Ross program, I was introduced to mathematical logic, through Father Ivo Thomas, visiting from Notre Dame, and one of the counselors, John P. Burgess, who is now a well-known logician and philosopher of mathematics at Princeton. During those summers, we went over Gödel's completeness and incompleteness theorems in some detail. We were also introduced, in general terms, to issues in modal logic, axiomatic set theory, and category theory. Philosophy was also addressed in the program, mostly in informal evening (and, sometimes, all night) discussions with faculty, counselors, and fellow students.

I thus arrived at Case Western Reserve University, as a freshman, knowing that I wanted to study logic and the foundations of mathematics, and—to the despair of my parents—not particularly thinking about or caring about what sort of career that would train me for. I soon declared a major in mathematics, devouring all of the foundational courses on offer, from the departments of mathematics, philosophy, and computer science. Along the way, I developed an interest in philosophy, and eventually took on a second major. After getting advice from the logicians at CWRU, I applied for and was admitted to the Ph.D. program, in mathematics, at the State University of New York at Buffalo. There, I learned from such logicians as John Myhill, Nicholas Goodman, Harvey Friedman, Thomas Jech, Leo Harrington, Richard Vesley, Akiko Kino, John Case, John Corcoran, and John Kearns. In addition to the staples of model theory, proof theory, set theory, and recursion theory, I was introduced to intuitionism and constructivism. Along the way, I got to see, more or less first hand, that there is controversy in mathematics. There is room for rational doubt and persuasion that goes beyond deductive reasoning using agreed-upon rules and axioms. In retrospect, this was the first challenge to what seemed obvious to my earlier self, namely that mathematics is the paradigm of certainty, the one place where all possibility of rational doubt is removed.

There was a wonderful spirit of cooperation between the logicians in the mathematics and philosophy departments, as well as the logically-minded computer scientists, at Buffalo. Toward the end of my first year, I discovered that my interests were turning more philosophical, and, at the same time, I was beginning to feel confident in my abilities in philosophy. I took a Masters degree in mathematics and transferred to the Philosophy Department. In retrospect, one main reason for the switch was that the philosophy program allowed me to pursue my interests in mathematics, while the mathematics program did not encourage such "outside" interests. I was also finally thinking about a career, about life after graduate school. I thought that I would enjoy teaching general and introductory philosophy more than I would enjoy teaching general and introductory mathematics.

Much of my work as a graduate student centered on the notion of computability. With Church's thesis as an example, I began thinking about the relationship between rigorously defined mathematical (or scientific) notions and their informal or pre-theoretic counterparts. The study of computability and constructivity also raised issues concerning intensionality in mathematics. A number of the intuitive or pre-theoretic notions studied in mathematical logic, such as computability, provability, and definability are at least prima facie intensional. At least intuitively, a function is computable only under a given description. That is, a function is computable under some descriptions and not others. Yet the usual treatments of these notions are thoroughly extensional. An intensional mathematics might allow one to better model the pre-theoretic notions.

My original interest with computability also led me to ponder the relationship between mathematics and physical reality. The problem was especially pressing for me, since one of the dominant views at Buffalo was platonism in the philosophy of mathematics. How is it that studying an ideal, unchanging realm of abstract objects is necessary to come to a deep understanding of physical reality?

There was no single dominant view in Buffalo, however. I witnessed much deep discussion of difficult and challenging problems, and soon started participating. I also benefitted from many discussions and presentations at the Buffalo logic colloquium. Two items of note are the debate over the status of second-order logic, which was quite intense, at least informally, there, and the emergence of structuralism in the philosophy of mathematics. I was inspired by the dissertation of, and discussion with my fellow student, the late Jane Terry Nutter.

What examples from your work illustrate the use of mathematics for philosophy?

My 1997 book, *Philosophy of mathematics: structure and ontology*, articulates a view that I call ante rem structuralism. The slogan of structuralism is that mathematics is the science of structure. The "ante rem" prefix denotes the metaphysical view that structures exist independently of any instances they may have. My book, as well as a related one by Michael Resnik, contains considerable metaphysics, ontology, epistemology, and an account of applications, but not much mathematics. At the urging of some colleagues and critics, I did include an axiomatization of structure-theory. That may count as mathematics, but even so, it is not very deep. The formal theory is modeled after Zermelo-Fraenkel set theory.

My other main interest in the philosophy of mathematics is the defense of second-order logic, as in my 1991 book, *Foundations without foundationalism: a case for second-order logic*, for example. That work does involve some mathematics, although it is not particularly difficult. To argue that the languages of mathematics are best (or at least well) interpreted as second-order, I provided reconstructions of common mathematical notions, inferences, etc., and provided careful comparisons of the first- and second-order formulations of various notions and theories. In exploring the expressive resources of second-order languages, one is led to its model theory, presenting standard and Henkin semantics, along with explorations of categoricity, Löwenheim-Skolem properties, compactness, indescribable cardinals, Lindström's theorem, etc. And, of course, second-order logic takes on a life of its own, with its distinctive mathematical properties. I have also contributed, in a modest way, to the industry of exploring intermediate logics, those that are not as expressively impoverished as first-order logic while not as intractable as full second-order logic. That involves working out the syntax, model theory, and sometimes, proof theory of the formal languages, and comparing those to the more usual first- and higher-order systems.

As noted above, early in my career, I got interested in intensional notions, and, with that, the interaction of constructive and non-constructive reasoning in mathematics. That led me to develop a classical formal system with a modal operator that could interpret the intuitionistic connectives and quantifiers. The language can capture statements in the form: constructively, there exists a number n, such that, classically, ... The solution to Post's problem is of this form. I stuck to arithmetic, and was able to establish that the translation of intuitionistic arithmetic is faithful. The project caught the interest of some of my teachers at Buffalo, Nicholas Goodman and John Myhill in particular, who proved the completeness of the translations, and extended the project to set theory, as well as stimulating some of the mathematics graduate students to contribute.

A bit more recently, I have taken an interest, as a sort of outsider mechanic, in the Scottish neo-logicist program in the philosophy of mathematics, sometimes called abstractionism. The idea is to develop branches of mathematics from abstraction principles, in the form

$$\text{(ABS)} \quad \S a = \S b \equiv \Phi(a,b),$$

where a and b are variables of a given type, typically first-order or second-order, $\S$ is a function from items of the given type to objects, and Φ is an equivalence relation of items of the given type. The program began its life with a principle about cardinal numbers now called Hume's principle:

$$\text{(HP)} \quad \#P = \#Q \equiv (P \sim Q),$$

where $(P \sim Q)$ is the statement that there is a one-to-one relation from the P's onto the Q's. Frege [1884] contains the essentials of a derivation of the basic principles of arithmetic, the so-called Dedekind-Peano axioms, from Hume's principle. The result is now known as Frege's theorem.

Mathematics itself enters the program in several ways. The abstractionist is keen to develop more advanced branches of mathematics from abstraction principles. This involves mathematical work, akin to Frege's theorem, although some of it, at least, consists of adapting well-known results to the framework. I have presented and discussed one such attempt for real analysis [2000] and discussed the prospects for set theory [2003].

Mathematical results also bear on the abstractionist program through the so-called "bad company" objection. The argument is that one cannot claim that Hume's principle has the requisite epistemological status since it is of the same form as another abstraction principle, Frege's Basic Law V,

$$\text{(BLV)} \quad \mathbf{E}P = \mathbf{E}Q \equiv \forall x(Px \equiv Qx),$$

which, of course, is inconsistent.

The main abstractionist response is to formulate properties, such as various conservativeness constraints, that distinguish the good abstraction principles – the ones that found branches of mathematics – from the bad ones. Most of the proposed conditions have mathematical content, and so abstraction principles can be adjudicated, in part, along mathematical lines. Alan Weir and I [1999] pointed out that a proposed abstraction for set theory, George Boolos's New V, fails the proper formulation of conservativeness, since it entails that the universe is well-ordered. We also show that it is consistent with ZFC that New V has no uncountable models (in which case, of course, it cannot found real analysis). In another contribution [2000], Weir and I examine the semantic and proof-theoretic properties of the logic used in the abstractionist program.

Some of the mathematical results show that under certain assumptions, the abstractionist program makes demands that are independent of, and sometimes incompatible with, the established model theory based on Zermelo-Fraenkel set theory. The relevance of these results depend on the attitude that the abstractionist takes toward previously established mathematics. This, in turn, raises questions concerning the relationship between philosophy of mathematics, and mathematics itself, which leads to the next question.

What is the proper role of philosophy of mathematics in relation to logic, foundations of mathematics, the traditional core areas of mathematics, and science?

Some of the examples in the previous section indicate that mathematical results can bear on, and help adjudicate and direct, philosophical programs. The question here concerns traffic in the opposite direction, from philosophy to mathematics, and to its logic. I have written on this topic, or at least parts of it, elsewhere ([1994], [1997, Chapter 1, §1]), and so I will be brief here. One view, with a long history, is that philosophical matters determine the proper practice of mathematics. One describes or discovers what mathematics is all about—whether, for example, there are mathematical entities that exist independent of the mind, community, or form of life, of the mathematician. This fixes the way mathematics is to be done. Call this the *philosophy-first* perspective. Of course, the order here is not historical, suggesting that mathematicians themselves first engage in philosophy, or consult their philosophical colleagues. The order is conceptual, or metaphysical, aimed at the proper foundational hierarchy.

If the first philosopher finds that mathematics is not done according to the prescribed canons, then she demands or argues that the practice be changed to conform to said canons. This is *revisionism*. Objections to the law of excluded middle, impredicative definitions, the axiom of choice, arbitrary and extensional functions and sets, and other principles and inferences, have been brought on philosophical grounds.

Although philosophy-first is alive today, at least in approximation, it is not prominent among philosophers of mathematics. Some of us even go to the opposite extreme, holding that philosophy is irrelevant to mathematics and perhaps to science as well. A position in the philosophy of mathematics is at best an epiphenomenon which has nothing to contribute to mathematics, and at worst a meaningless sophistry, the rambling and meddling of outsiders. I have called this the *philosophy-last-if-at-all* perspective. It leads to a sort of quietism concerning mathematics.

Stark versions of either of these extreme perspectives presuppose, or seem to presuppose, a sharp distinction between mathematics (or logic or science) and its philosophy. To be sure, a lot of good mathematics has been inspired by the philosophical orientation of the mathematician. On the other hand, a given philosophical perspective may lead to the suppression of good mathematics. Suppose, for example, that one is convinced of the universal validity of the law of excluded middle, say on the basis of a priori rational intuition. Then one will reject, out of hand, those branches of mathematics, such as intuitionistic analysis and smooth infinitesimal analysis, that would collapse into nonsense if excluded middle were enforced.

The wise course, I submit, is to wax holistic and think of mathematics and its philosophy as overlapping but roughly autonomous disciplines, with neither dominating the other. There is cross-fertilization, but the mathematician and the philosopher are asking different questions, and their methodologies are different as well. I confess to having no argument in support of this holistic position. Indeed, it is hard to see what such an argument would look like. What would be its premises?

I take the main goal of the philosophy of mathematics to be to *interpret* mathematics, and thereby answer philosophical questions concerning the place of mathematics in the world view. One focus is on the language of mathematics. What do mathematical assertions mean? What is their logical form? What is the best semantics for mathematical language? How is mathematical language to be understood? And there are more global questions. What is the subject matter of mathematics—if it has one? What is the relationship between mathematics and the subject matter of science which allows such extensive applications? How do we manage to do and know mathematics? How can mathematics be taught? A complete world view would include something about mathematics, something about the applications of mathematics, something about mathematical language, and something about ourselves, showing how all of these aspects interact with each other.

This perspective runs against the revisionism that sometimes goes with philosophy-first. It is *mathematics* that is to be interpreted, and not what a prior (or a priori) philosophical theory says that mathematics should be. To be sure, interpretation can and should involve criticism. No one is infallible, and anyone can point out mistakes in a given practice. But if a philosopher has the practitioner making too many mistakes, it is perhaps reasonable to conclude that he is changing the subject. Perhaps it is the philosophy and not the mathematics that is at fault.

As I conceive it, philosophy of mathematics is done by those who care about mathematics and want to understand its role in the intellectual enterprise, in the ship of Neurath. Ideally, a mathematician who adopts a philosophy of mathematics should benefit from it, at least if the philosophy is plausible. She will obtain an orientation toward her work, and some insight into its perspective and role.

What do you consider the most neglected topics and/or contributions in late 20th century philosophy of mathematics?

What are the most important open problems in the philosophy of mathematics and what are the prospects for progress?

With these final questions, I must get more speculative, and even more tentative. There is also some modesty and self-criticism involved in answering this question. If I find a given topic neglected, then, it seems, I myself neglected it. And if I find a problem open, then I have not found a resolution that satisfies me.

I must confess that when I hear or read a philosopher, mathematician, or historian complaining that philosophers of mathematics neglect something or other, I tend to agree. We have been told, for example, that philosophers do not pay enough attention to the details of various branches of mathematics. We do tend to focus exclusively on either the most elementary parts, such as arithmetic, or on the more foundational branches, such as set theory or perhaps category theory. We also hear that philosophers show ignorance of important themes in the history of mathematics. Or that philosophers of mathematics are unaware of the more subtle ways that mathematics finds application in the various sciences, either in the inner reaches of physics, or in the statistics of psychology and economics, or someplace in between.

The charges are not completely fair. There are interesting and important questions about the natural numbers, and their everyday applications, that can be addressed without knowing any substantial number theory, let alone any functional analysis. Indeed, one can be led into distortion or outright error by paying too much attention to the more advanced areas of mathematics, and ignoring more mundane issues and applications. Moreover, some of the philosophical issues have cleaner formulations when one thinks about simpler instances of them. We have to figure out how we know the Dedekind-Peano axioms – if in fact we do – before worrying about whether we know or how we know the axioms and theorems of algebraic topology. We have to see how arithmetic applies before we turn to applications of analytic function theory.

Nevertheless, many of the charges are legitimate. Our focus often is too narrow. However, not many philosophers are in position to address the charges. To say something about the relevance of, say, homotopy theory, one must know something about homotopy theory. To address issues concerning the history of mathematics, one must know something about this history, and to address substantial and non-trivial applications, one must know something about the fields to which mathematics is applied—quantum field theory and economics, for example. More and more young philosophers of mathematics are getting training in the various disciplines, at least at a beginning graduate level, and some even come to philosophy from the various disciplines. So perhaps at least some of these shortcomings may be overcome in due course.

Another neglected area, which can be addressed by just about anyone in our discipline, is the relationship between philosophy of mathematics and other areas of philosophy. Consider, for example, the widespread (but not universal) rejection of foundationalist epistemologies. Do the relevant arguments have any bearing on foundationalism in mathematics? Or vice versa. Do foundationalist consideration in mathematics – such as they are – have any ramifications for general epistemology? Suppose that one decides that mathematics is an exception to a general principle in epistemology. What does that say about mathematics, and about the general principle?

To take another example, "externalism" is a name for a variety of views that do not require the justification for a known proposition to be available, internally, to the knower, or to anyone else for that matter. Arguably, the standard of justification in mathematics is proof. Are various forms of externalism plausible for mathematics? Or is mathematics different from ordinary empirical knowledge in this respect? One item related to externalism is the question of when a belief-producing faculty is reliable. How does this issue play itself out with respect to mathematics? What are the belief-producing faculties in mathematics, and what does it mean for them to be reliable?

Another area to be mined is the philosophy of language. What happens when the most plausible semantic accounts are applied to the language of mathematics? How do treatments of singular terms generally bear on the singular terms of mathematics (if there are any)? What of predicates? Is some sort of contextualism plausible for mathematics? What are the relevant contexts? Suppose that contextualism is found to be a non-starter with respect to mathematics. Does this undermine its plausibility elsewhere, or is there some insight about the nature of mathematics to be found by examining the differences?

The problem of indistinguishable participants concerns cases where it seems to be impossible to identify the referent of a pronoun or other singular term. One classic case is the sentence:

> When a bishop meets another bishop, he blesses him.

The problem of formulating truth conditions for such sentences is, it seems, interesting. Mathematics provides some rather clean examples of indistinguishable participants. In complex analysis, for example, it can be shown that the symmetry between the two square roots of -1 (or any other complex square roots, for that matter) cannot be broken. Yet the language of complex analysis has something that at least looks like a singular term that denotes one of them, namely 'i'. Which of the square roots of -1 does 'i' denote? Similarly, mathematicians use the definite description construction, talking about, say, "the square root of -4". Can that be handled along the lines of other definite descriptions, elsewhere? Surely, there are lessons to be learned about how singular terms are deployed, and how languages are learned, here.

In some branches of elementary mathematics, such as Euclidean geometry, *all* of the objects are indistinguishable. Yet like any other language, geometry requires the use of pronouns, or at least seems to. Does this say anything about language in general?

One area of general neglect in general philosophy is the notion of explanation. We speak of explanations all the time, both in ordinary discourse and in serious philosophy. Sometimes a philosopher takes an inference to the best explanation, or even a simple claim that she has explained something, as a guide to truth in some area of philosophy. And we praise our pet theories as providing explanations of various intuitions, or various semantic constructions, or whatever. But what is it to explain something? Intuitively, to explain is to clear something up, to remove puzzlement or obscurity. That seems like a subjective notion, tied to one's interests at a given time. A chemist might be satisfied with the explanation that a fire occurred because of the combustion of exposed gasoline, but an arson inspector may not be not satisfied with this same explanation. For the latter, the proposed explanation does not clear up what was confusing, or at least what was of interest.

The one area of philosophy that has paid considerable attention to explanation is the philosophy of science, with its extensive literature and ongoing debates on this topic. But at least some of the accounts of explanation there are not appropriate outside of science. For example, theories that turn on causal matters do not make sense of – or explain – explanations in non-causal areas of discourse. Moreover, the extent to which scientific explanation is an instance of explanation generally remains unclear, or at least unproven.

Mathematics provides an interesting and perhaps central case study for this problem. Mathematicians regularly speak of explanatory proofs, explanatory theorems, explanatory examples, and the like. There has been some work on mathematical explanation, especially lately. A growing literature examines cases that mathematicians find explanatory, relating those to the various accounts of explanation in science and the few accounts proposed for explanation in mathematics—usually with negative consequences concerning the various accounts. I hope, and expect, this work to continue, resulting in a positive account of mathematical explanation. This, perhaps, will go some way toward illuminating the general notion of explanation.

Turning to our final, closely related example, Georg Kreisel is reputed to have said that the important question in philosophy of mathematics is not the existence of mathematical objects, but the objectivity of mathematical discourse. This raises a fundamental question concerning the nature of objectivity. What is it for something, or some assertion, to be objective?

It is a truism that features of scientific theories are a function of the way the world is and the way humans are and, in particular, the ways that humans apprehend the world, both as individuals and socially. Philosophers as diverse as Immanuel Kant and W.V.O. Quine have argued that it is impossible to sharply separate out these features of our theories and discourses. It is not that the world itself is a human creation, as some have claimed. The claim here is only that our *theories* of the world are human creations, at least in part. What is added to this truism is a claim that, to wax metaphorical, it is not possible to cleanly separate out the "world" part from the "human" part.

This perspective, if correct, calls the very notion of objectivity into question. Intuitively, to be objective is to be independent of human judgements, sensibilities, languages, communities, forms of life, or whatever. The Kant-Quine orientation would at least suggest that there is no objectivity, or at least no pure objectivity, to be had. There simply is no perspective available to us which is independent of human judgements, perspectives, etc. To switch to another metaphor, there is no God's eye view that we can take, in order to compare our theories and judgements with objective reality.

Nevertheless, the notion of objectivity is an important one in ordinary discourse, and in our intellectual lives. There at least seems to be a clear difference between statements like "Smoking is cool", "Broccoli is disgusting", "George Carlin is funny", and "The skater put on a brilliant show", and statements like "smoking causes cancer", "the acceleration due to gravity near the earth is 9.8 meters per second per second", "the runner completed the race in under four minutes", "7+9=16", and "there are arbitrarily large prime numbers". But what is this distinction? How is it to be made out?

An advocate of the Kant-Quine orientation may conclude that there simply is no notion of objectivity. Every statement that a given proposition or judgement is objective is false. Perhaps an error theory is called for. An advocate of this view can either argue that the notion of objectivity should be dropped from our vocabulary, having it go the way of phlogiston. Or the error theory might be supplemented with an account of the social or conversational role of attributions of objectivity. But I, at least, hope we can do better. An error theory has to be a last resort, embraced only when we despair of a more attractive account.

I speculate that mathematics is an important case study in the quest for a good account of objectivity. It sure seems like mathematical discourse is objective, in that, for example, the correctness of calculations or purported proofs does not depend on human judgements. For one thing, mathematicians seem to come to consensus on purported proofs rather quickly, much quicker than in just about any other area of human inquiry—debates over intuitionism and relevance logic aside. Moreover, the consensus in the mathematical community is remarkably stable, at least at this point in history. I take it that this is at least some evidence that mathematics, and mathematicians, are tracking something objective—whatever that objective thing may be. To be sure, one can explore alternative explanations of the consensus in the mathematical community (whatever it is to be an explanation). But it at least seems worthwhile to attempt to articulate and defend the thesis that mathematics is objective. And this requires a more detailed account of what it is to be objective.

References

Frege, G. [1884], *Die Grundlagen der Arithmetik*, Breslau, Koebner; *The foundations of arithmetic*, translated by J. Austin, second edition, New York, Harper, 1960.

Shapiro, S. (editor) [1985], *Intensional mathematics*, Amsterdam, North Holland Publishing Company.

Shapiro, S. [1994], "Mathematics and philosophy of mathematics", *Philosophia Mathematica (3) 2*, 148–160.

Shapiro, S. [1997], *Philosophy of mathematics: structure and ontology*, New York, Oxford University Press.

Shapiro, S. [2000], "Frege meets Dedekind: a neo-logicist treatment of real analysis", *Notre Dame Journal of Formal Logic 41*, 335–364.

Shapiro, S. [2003], "Prolegomenon to any future neo-logicist set theory: abstraction and indefinite extensibility", *British Journal for the Philosophy of Science 54*, 59–91.

Shapiro, S. and A. Weir [1999], "New V, ZF, and abstraction", *Philosophia Mathematica (3) 7*, 293–321.

Shapiro, S. and A. Weir [2000], "'Neo-logicist' logic is not epistemically innocent", *Philosophia Mathematica (3) 8*, 163–189.

22

Wilfried Sieg

Professor of Philosophy
Department of Philosophy
Carnegie Mellon University, USA

Philosophy of Mathematics: 5 Questions, 4 Answers and 4 Problems

Why were you initially drawn to the foundations of mathematics and/or the philosophy of mathematics?

When Berlin was still divided, I started studying mathematics and physics at the Technical University in Charlottenburg. The labs for experimental physics were a wonderful adventure, and constructing precise drawings for "Darstellende Geometrie" was a real pleasure; but my mathematics education, I felt, was far too applied and too much directed towards engineers. So I decided to enroll at the Free University in Dahlem. The difference could not have been more striking: no labs in physics, but a great course drawn from Richard Feynman's "Lectures on Physics"; no constructing of intersection curves, but rather chasing of diagrams in an introduction to category theory. Peter Grotemeyer taught that course under the title "Lineare Algebra." The atmosphere at the Second Mathematical Institute, directed by Grotemeyer, was open and exciting; Bourbaki as well as MacLane and Eilenberg had shaped the intellectual outlook. The tension between this approach to mathematics and a more classical one was palpable, as Alexander Dinghas – the Director of the First Mathematical Institute – vividly represented the latter. The rather open conflict pushed me to read more on foundational topics, as no courses on such issues were offered. I worked through standard texts in logic and set theory, but read also many essays by Paul Lorenzen and his book "Operative Logik und Mathematik."

After my "Vordiplom" I went back to West Germany and continued studying mathematics at the Wilhelms-University in Münster, replacing physics by logic as my secondary field. This was easy and my reason for going to Münster, as the *Institute for Mathematical Logic and Foundational Research*, founded by Heinrich Scholz in the 1930s, was part of the Mathematics Department. Dieter Rödding directed the Institute at the time, and I received my basic education in logic through his engaging lectures on predicate logic, set theory, and recursion theory. A switch to foundational studies had been made, and I even started to take philosophy courses. I distinctly remember a seminar on the philosophy of mathematics in which a good deal of attention was devoted to Hilbert's Program; Friedrich Kaulbach offered that seminar. With some of my fellow students in the Institute I studied (with great admiration, but little understanding) Georg Kreisel's "Survey of Proof Theory" that had recently been published in the *Journal of Symbolic Logic* and also Solomon Feferman's more accessible "Arithmetization of Metamathematics".

It is perhaps not too surprising that I chose to go to Stanford, when I had the opportunity to spend a year abroad after completing my "Diplom" in mathematics; a fellowship from the DAAD allowed me to do so as a Visiting Scholar. At the end of that year I was accepted into Stanford's Ph.D. Program in Logic and Philosophy of Science. Five years later I had finished a dissertation, under Feferman's thoughtful and gentle guidance. My thesis "Trees in metamathematics (Theories of inductive definitions and subsystems of analysis)" contributed to proof theory by establishing the consistency of impredicative subsystems of classical analysis relative to intuitionistic theories of constructive number classes. During my time at Stanford I learned to balance detailed mathematical work in logic, especially in and related to proof theory, with broader philosophical reflections, and with my curiosity about the emergence of modern logic and foundational investigations.

What examples from your work illustrate the use of mathematics for philosophy?

In the history of philosophy one finds many uses of mathematics for philosophy. Here is a traditional example that consists of two conflicting parts: (i) Kant supported his claim that there are synthetic a priori judgments by pointing to mathematics, in particular also to arithmetic; (ii) Frege argued that his systematic development of arithmetic from logical principles establishes the analytic character of arithmetic. The second part involves a different understanding of use for philosophy than the first, an understanding I actually prefer; it appeals to mathematical and logical notions as well as techniques in order to sharpen and, perhaps, solve philosophical problems.

Mathematical logic is for me a part of mathematics. Thus, I answer this question in a way that is focused on its role in the foundations of mathematics. My first example sketches one strand of a *systematic inquiry* to which many logicians, mathematicians, and philosophers have contributed. It is characterized by the dynamic interaction of detailed mathematical work, strategic logical approaches, and broad reflections on the nature of mathematics. Connected with this inquiry is an important conceptual issue, namely, a rigorous characterization of "formality" or "computability." Building on work by Turing and Gandy, I have addressed this deeply methodological issue in a way that is a proper use of mathematics for philosophy; this will be my second example.

2.1 First example. The systematic inquiry takes on a problem that can be traced to the central foundational issue in 19th century mathematics, namely, securing a basis for analysis. A resolution of the problem is indicated by the slogan "Arithmetize analysis!" This directive was already given by Gauss, and its meaning can be fathomed from Dirichlet's claim that *any* theorem of analysis can be formulated as a theorem concerning natural numbers. For some the arithmetization of analysis was accomplished by the work of Cantor, Dedekind, and Weierstrass. For others a stricter arithmetization was required, one that does not appeal to set theoretic notions. Kronecker in particular rejected the general concept of irrational numbers, recognizing as legitimate only those that can be presented via effective approximations. Thus, the irrational numbers constituted an open-ended domain for him, whereas for Dedekind in (1872) it was crucial to determine their full extension as a gapless "continuous domain" by axioms and to provide the set of all cuts of rationals as a model.

In his (1888) Dedekind characterized the natural numbers via his axioms for simply infinite systems. For excellent methodological reasons articulated in his letter to Keferstein, Dedekind provided also here a model, i.e., an example of a simply infinite system. (As is well known, that particular example was obtained by using the inconsistent system of everything thinkable as a starting-point.) Hilbert followed Dedekind in the way he formulated axioms for geometry (1899) and for arithmetic (1900), namely, as conditions on assumed systems of things. Deeply influenced by both Dedekind and Hilbert, Zermelo saw his axioms for set theory in (1908) in the Dedekindian way as pertaining to a domain of individuals that are "subject to the following axioms, or postulates." This feature was emphasized, when Hilbert and Bernays referred to this early version of the axiomatic method as *existential axiomatics*. The difference to the later *formal axiomatics* is striking and has to be clearly seen.

In these early axiomatic developments neither a formal language nor a logical calculus was specified. That is of course the reason why Frege polemically claimed that there were no proofs in Dedekind's (1888). Hilbert asked in his (1905) for a simultaneous development of arithmetic and logic, but carried it out in an adequate way only in lectures for the winter term 1917/18; then he was deeply influenced by Whitehead and Russell's *Principia Mathematica*. In these lectures and those for the winter term 1921/22, Hilbert and Bernays presented the beginnings of analysis in a version of the system of *Principia Mathematica* with the axiom of reducibility, i.e., in full second order arithmetic (CA).[1] The formal theories introduced for logic and mathematics were also considered as objects of metamathematical studies, in complete and quite conscious analogy to the treatment of theories for geometry in Hilbert's *Grundlagen der Geometrie*.

However, these newly introduced formalisms allowed something dramatically different. For Dedekind and also Hilbert around 1900 consistency was a semantic notion and had to be established by giving an arithmetico-logical example of a system satisfying the axiomatic conditions. In early 1922 proof theory was invented: exploiting the elementary character of the description of formal theories, Hilbert noticed that they can be defined in finitist mathematics and that the consistency problem in particular can be formulated there; so, why shouldn't it be possible to also solve it with finitist means? The answer to this question was given, not

---

[1] This was the basis for the beautiful presentation of classical analysis in the second volume of *Grundlagen der Mathematik*, Supplement IV. The presentation in *Grundlagen* was, on the one hand, influenced by Weyl's *Das Kontinuum* and, on the other hand, served as the basis for refinements in the context of proof theoretic investigations in the 1960s and 1970s. The later work was the starting point for "Reverse mathematics" investigating which theorems of particular parts of mathematics are equivalent to set existence principles in subsystems of analysis.

even a decade later, by Gödel's second incompleteness theorem, limiting rather severely the possibility of consistency proofs.

A generalized reductive program in Hilbert's spirit, but with a broadened constructive base has been pursued ever since Gödel and Gentzen established the consistency of classical elementary arithmetic relative to its intuitionistic version. The first important proof theoretic result, which went beyond Herbrand's for elementary arithmetic with just quantifier-free induction, was Gentzen's 1936 consistency proof for the full system. Dramatic progress has been made since then in two complementary directions. On the one hand, the core of classical analysis has been carried out in theories that are conservative over elementary arithmetic. On the other hand, strong impredicative subsystems of (CA) have been reduced to constructively acceptable theories of generalized inductive definitions, in particular, constructive higher number classes. These results lead naturally to a programmatic distinction between two types of reductions: *Foundational reductions* are to provide a constructive basis for strong and, from particular perspectives, problematic parts of (CA); *computational reductions* are to yield algorithmic information from proofs in weak and, even from a finitist standpoint, unproblematic parts of (CA).

Foundational reductions push forward the generalized program and attempt to answer the question "What more than finitist mathematics do we have to know to recognize the soundness of a strong theory?" In sharp contrast to the original Hilbert program, this reductive program is not seen as an instrument to solve the foundational problems of mathematics once and for all; it is rather viewed as a coherent scheme guiding foundational research. Work in its pursuit has given us a much clearer understanding of the mathematical strength of classical theories, the relation between classical and constructive theories, and the nature of constructive principles and their relative strength. But reductive results do not only provide information; they also present a philosophical challenge. After all, the reductions do secure classical theories on the basis of theories, which can justly be called more elementary. In this sense they are epistemological reductions, whose precise character awaits philosophical analysis.

2.2 Second example. The above discussion takes for granted that we are dealing with "formal" theories, in particular when articulating Hilbert's Program and the effect the incompleteness theorems have on it. Only when we have a mathematical characterization of "formality" can Gödel's theorems be stated as pertaining to *all* formal theories; only then can we formulate a "completely general version" of these theorems. "That is", as Gödel asserted in a 1963 note added to his (1931), "it can be proved rigorously that in *every* consistent formal system that contains a certain amount of finitary number theory there exist undecidable arithmetic propositions and that, moreover, the consistency of any such system cannot be proved in the system." Such a characterization appeals first to informal concepts like "mechanical procedure", "finite combinatory process", or "effective calculability"; these informal concepts are then "identified" via a generic form of Church's Thesis with a precise mathematical concept.

The first published version of Church's Thesis was formulated for effectively calculable number theoretic functions; in (1935) Church suggested identifying them with Gödel's general recursive functions. The reasons hinted at in Church's brief abstract and then expanded in his (1936) are still central when arguments for the thesis are presented: (i) the quasi-empirical observation that all known calculable functions can be shown to be general recursive, and (ii) the mathematical fact of the equivalence of differently motivated notions. (In (1935) Church referred to the equivalence of λ-definability and general recursiveness.) The support for the thesis through the arguments *from coverage* and *confluence* was deepened in (Church 1936) by the step-by-step argument *from a core conception*. Church explicated the effective calculability of a function by its "calculability in a logic" using elementary steps; these steps were taken by Church to be general recursive ones. With that "central thesis" concerning steps in place, effectively calculable functions are indeed general recursive.

All these arguments are in the end deeply unsatisfactory. The quasi-empirical observation could be refuted tomorrow, as we might discover a function that is calculable, but not general recursive. The mathematical fact by itself is not convincing, as the ease with which one can prove the equivalences shows a deep family resemblance of the different notions. The explication of effective calculability via the core concept "calculability in a logic" is important and achieves a conceptual sharpening; however, a mathematically convincing result is obtained only by an appeal to the central thesis, i.e., semi-circularly. Gödel's equational calculus is the paradigmatic "logic" for Church, clearly in a very special and restricted setting. For Gödel the equational calculus seems to have provided a "canonical" way of analyzing the calculation of values of (recursively specified) functions. In addition, Gödel observed in 1936 the "absoluteness" of the notion of general recursive functions, but formulated it in (1946) in a more general way: if a function is calculable in a formal theory extending elementary arithmetic, then it is calculable already in elementary arithmetic.

Gödel must have realized that any argument for the metamathematical fact of absoluteness must exploit in some form the "formal" character of the stronger theories (and thus make a similar semi-circular move as I pointed out in Church's argument for the recursiveness of effectively calculable functions). In 1964, Gödel wrote a Postscriptum to the 1934 Princeton Lectures, extending the note he had written a year earlier for his 1931 paper. He re-asserts the claim that Turing's work provides the basis for "a precise and unquestionably adequate definition of the general concept of formal system." But in addition, he states a reason why Turing's work does so: "Turing's work gives an analysis of the concept of 'mechanical procedure' (alias 'algorithm' or 'computation procedure' or 'finite combinatorial procedure'). This concept is shown to be equivalent with that of a 'Turing machine'."

Gödel did not explain this very schematic remark on Turing's work. However, Turing's analysis of *human* computability does lead very naturally to boundedness and locality restrictions; they in turn motivate axiomatic conditions on a very general class of discrete dynamical systems; an appropriate representation theorem asserts that any model of the axioms is computationally reducible to a Turing machine. In my (2007) one finds a brief exposition of the crucial issues and references to detailed conceptual and mathematical work. The considerations I sketched all too briefly do not appeal anywhere to a thesis or even a central thesis; they are straightforwardly mathematical—up to a point. The latter qualification is needed, as they do appeal for the formulation of the axiomatic conditions to an informal understanding of notions, here of human computability. Such an appeal can no more be avoided in this case than in any other case of an axiomatically characterized structure intended to model aspects of physical or intellectual reality. In short, we do not have to face any "mysterious" and special issue surrounding the concept of effective calculability. Rather, we have to face the ordinary issues of judging the adequacy of mathematical concepts for particular goals. Of course, these issues are far from trivial. As in the first example, the mathematical work leads to a clarification and deepening of the relevant philosophical questions.

What is the proper role of philosophy of mathematics in relation to logic, foundations of mathematics, the traditional core areas of mathematics, and science?

What do you consider the most neglected topics and/or contributions in late 20th century philosophy of mathematics?

The extended discussion concerning question 2 reflects what I view as the proper role of philosophy of mathematics in relation to logic, foundations of mathematics, and the core areas of mathematics. Nevertheless, let me emphasize that, for me, philosophy of mathematics is so deeply intertwined with logic and the foundations of mathematics, that I don't know how to cleanly separate them – if the latter are pursued with a reflective outlook. As to the core areas, they should be accounted for and not be dismissed, in case they don't fit with a restrictive philosophical position; after all, they are flourishing and proceed in general quite independently of foundational investigations.

One aspect of the relation to science is implicit in my earlier discussion: the development of analysis in weak theories undermines indispensability arguments for set theory. However, mathematics has a strong constitutive role for scientific investigations through providing sharp models of reality. The thorough examination of the various components involved in this formulation should be a joint enterprise for philosophy of mathematics and philosophy of science and, perhaps, lead back to the balance and integration that was prevailing in the early part of the 20th century, for example in the work of Hilbert, Poincaré, and Weyl. Weyl wrote his contribution to the *Handbuch der Philosophie* (1928) on "Philosophie der Mathematik und Naturwissenschaft." In a contemporary version of such a contribution, one would have to treat not only the natural sciences, but also computer science and cognitive science, in particular, cognitive psychology. The integration of mathematics and science is for me the most neglected topic in late 20th century philosophy of mathematics.

The remarks on the "most neglected topic" are to be complemented with remarks about "the most neglected contribution;" well, what I have in mind is not exactly "the" most neglected contribution in "late" 20th century philosophy of mathematics, but rather a long sequence of important contributions that have been made over the second and third quarter of the last century. I am thinking of the work of Paul Bernays, but only in part of the monumental *Grundlagen der Mathematik*, the two-volume work he wrote and published with Hilbert's imprimatur in 1934 and 1939; I am mostly thinking of the specifically philosophical essays he composed between the mid-twenties and the mid-seventies. Many of those essays were collected in his *Abhandlungen zur Philosophie der Mathematik* published in 1976 and contain, in my view, the most important and sustained reflections on mathematics in the 20th century. They are also a commentary on the foundational investigations during that time, as they incorporate most of the philosophically significant developments in some form or other.

As an example consider the long essay "Die Philosophie der Mathematik und die Hilbertsche Beweistheorie." The essay was written during the summer of 1930 just before Gödel's incompleteness theorems were discovered, but some of the detailed considerations go back to the early and mid-twenties. It presents a marvelous analysis of the finitist standpoint and of the state of proof theoretic investigations, but discusses also connections to developments in the second half of the 19th century and contrasts the finitist approach with other foundational perspectives. In particular, Bernays discusses Brouwer's intuitionism, which is not viewed as a different foundational perspective, but rather – quite usual at the time – as coextensional with finitism. (Bernays clearly articulates this view at the top of p. 42 and in footnote 9 attached to that text.) Brouwer's intuitionism is distinguished from Hilbert's finitism by a different philosophical motivation and, obviously, by not sharing the programmatic goal of securing analysis and set theory with full classical logic. Only a few years later this view is overthrown by the metamathematical results I mentioned already, namely, Gödel and Gentzen's consistency proof of classical elementary arithmetic relative to its intuitionistic version.

What are the most important open problems in the philosophy of mathematics and what are the prospects for progress?

My remarks here directly link up with the earlier ones regarding the first example for question 2, where I pointed to the philosophical challenge presented by the reductive results and the broader context in which they figure prominently. The results, it seems, ask us to reflect anew on the questions "What is to be reduced, and why?" and "What is considered to be basic, and on what grounds?" The fundamental informal idea underlying (Bernays' expositions of) Hilbert's finitist program is this: project *any* axiomatically characterized structure by means of the associated formal theory into the finitist domain, and recognize the consistency of the theory from the finitist standpoint. It is this uniform reduction to a single elementary standpoint that has to be given up on account of the second incompleteness theorem. The finitist domain, providing the objective underpinnings for the recognition of principles that are used in consistency proofs, has to be extended in a way that respects a broadly understood *constructive aspect* of mathematical experience.

This constructive aspect reflects our grasp of the uniform generation of mathematical objects, but also our understanding of proof and definition principles grounded in the generating processes. Here are examples of such objects: natural numbers, formulas of predicate logic, constructive ordinals, and sets in segments of the cumulative hierarchy. Aczel describes in his (1977) a broad framework for the inductive generation of mathematical objects. The rules, which underlie the generation of objects, are so general that they encompass all those for the examples I mentioned. Indeed, they can be taken to be *deterministic* in that objects have to be generated from unique sets of premises to become elements of *accessible domains*. The determinism guarantees the existence of *canonical isomorphisms* between accessible domains that are closed under the same rules.[2] Though *accessible domains* are structures, their elements are not featureless points, but have an important internal structure of their own that shows them to be in the accessible domain. I see a sharp contrast between accessible domains and structures that fall under notions like group, field, topological space and differentiable manifold. These abstract notions have strikingly different instances and are at the heart of the *conceptional aspect* of mathematical experience; they allow us to comprehend complex connections, to make analogies between different theories precise, and thus to obtain a more profound understanding. Relative consistency proofs have the task of establishing the coherence of these notions relative to accessible domains.

---

[2] These accessible domains differ only through their base elements, and permutations of these atoms uniquely determine the isomorphims. If we allow an abstractive move, what W.W. Tait calls "Dedekind abstraction" in the case of simply infinite systems, we seem to be justified in speaking of mathematical objects. In contrast to Dedekind and Tait, I do not think that the real numbers can be introduced by "Dedekind abstraction"; though any two complete ordered fields are indeed isomorphic, the isomorphism is not *canonical* in my sense.

Here is my *first question*: Along which theoretical dimensions can we understand the important role abstract notions play in mathematical practice?—Wittgenstein is reported as having said, "If you want to understand a theorem, look at its proof!" Proofs provide *explanations* of what they prove by putting their conclusions in a context that shows them to be correct. This is usually not a global context providing a foundation for all of mathematics, but a rather more restricted one. Indeed it can be quite local, when it just analyzes the crucial insights that are at the heart of an argument. Two informative, but perhaps slightly non-standard illustrations of what I have in mind can be found in *Grundlagen der Mathematik II*. The first illustration, from the second supplement to that volume, is provided by the "recursiveness conditions" and can be viewed as a precise mathematical formulation of Church's step-by-step argument. Number theoretic functions are considered as "reckonable" (regelrecht auswertbar) when their values can be calculated in quite open deductive formalisms. As soon as the proof relation of such a deductive formalism is required to be primitive recursive (the central recursiveness condition), one can prove a theorem: all reckonable functions are general recursive. The second illustration concerns a related case: the proof of Gödel's second incompleteness theorem is structured in such a way as to bring out the essential "derivability conditions" for formal theories. Such a *local deductive* organization seems to be the classical methodology of mathematics for gaining a "real" understanding of a theorem, and that may be gained from perspectives emphasizing different ways of looking at a problem, e.g., from a topological or algebraic point of view.

The task of considering a significant part of mathematics, finding appropriate basic notions, and explicitly formulating principles – so that the given part can be systematically developed – is of a quite different character. For Dedekind in (1854) the need to introduce new and more appropriate notions arises from the fact that human intellectual powers are imperfect. The limitation of these powers leads us, Dedekind argues, to frame the object of a science in different forms or different systems. To introduce a notion, "as a motive for shaping the systems," means in a certain sense to formulate a hypothesis concerning the inner nature of a science, and it is only the further development that determines the real value of such a notion by its greater or smaller *efficacy* (Wirksamkeit) in recognizing general truths. Dedekind put the axiom systems in both his foundational essays to this test; it was of the highest importance to him in (1872) that a central part of real analysis could be developed, and in (1888) that the standard part of elementary number theory could be obtained directly.

No matter how one might attempt to systematically gain such a more principled understanding of a part of mathematics, *local axiomatics* of the sort described may be directly tested; namely, by examining what else is needed, beyond a logical calculus and local axioms, for an efficient *automated* development.[3] Together with Clinton Field, I carried out such a test for the incompleteness theorems; see our (2005). Assuming as local axioms only the representability and derivability conditions (and the diagonal lemma), the automated theorem prover **AProS** very efficiently finds proofs of the theorems. The success of the search procedure results from carefully interweaving logical and mathematical considerations. In addition, these considerations are made dynamic by logical strategies and heuristic mathematical ideas; as the central heuristic idea we used the principled back and forth between arguments in the object and meta-language. Currently, I am testing this approach for an elementary part of set theory leading up from the axioms of Zermelo-Fraenkel to the Cantor Bernstein theorem; I am not quite sure yet, how this experiment is going to come out.

My *second question* is this: Can we obtain by detailed studies of actual mathematical practice a deeper understanding of the role of mathematical concepts? Can we make a genuine advance in isolating basic operations of the mind or, in other words, can we develop a cognitive psychology of proofs that reflect genuine logical and mathematical understanding? – In his marvelous paper (1988), Howard Stein analyzed the profound transformation of mathematics in the 19th century, in which Dedekind played a central role. He calls this transformation the "second birth" of mathematics, "its first having occurred among the ancient Greeks, say from the sixth to the fourth century B.C." He points, in the case of the Greek developments, to mathematics as a "systematic discipline with clearly defined concepts and with theorems rigorously demonstrated." In both cases he regards these developments "as a discovery of the capacity of the human mind, or of human thought – hence its tremendous importance for philosophy

---

[3] I am thinking of fully automated proof search, but also of interactive theorem proving, proof planning, or proof verification. There is no conflict or even sharp contrast between these methods for my considerations here, as in each case one has to adopt some "local axiomatics".

... ." My second question has consequently a second formulation; namely, can we better characterize this remarkable capacity of the human mind by giving a refined analysis of the rich body of mathematical knowledge that is systematically organized, but also structured for intelligibility and discovery?

In the meantime, before receiving answers to these "big" questions, we can address two additional and I think challenging questions of a more concrete character, but not among "the most important open problems in the philosophy of mathematics." I formulated them already in my (2002); they refer back to the discussion of accessible domains. My *third question* asks for an abstract mathematical description of such domains that highlights their distinctive features: Can one give a category theoretic characterization of accessible domains? - Accessible domains play an important foundational role and provide means for consistency proofs, whether syntactic, proof-theoretic or semantic, model-theoretic ones. I am hoping that detailed mathematical and philosophical analyses will lead to informative distinctions that concern generating operations and their iteration, but also fundamental deductive principles. That leads to my *fourth question*, which is rather natural: Can theories for accessible domains be given in such a form that their classical versions are uniformly reducible to their intuitionistic variants?

The reductive program that emerged from Hilbert's provides an important foundational perspective on mathematics. I have been calling the associated philosophical position *reductive structuralism*: it allows us to connect in a coherent way the two aspects of mathematical experience I distinguished earlier; it helps us to gain a better understanding of modern mathematics (or mathematics after its second birth) and its role in the sciences; it allows creative freedom for *constructions* and *abstract concepts*, where the former call for abstract analysis, the latter for constructed models.

References

Aczel, P.

1977 An introduction to inductive definitions; in: *Handbook of mathematicval logic* (J. Barwise, ed.), North-Holland, Amsterdam, 739–782.

Bernays, P.

1976 *Abhandlungen zur Philosophie der Mathematik*; Wissenschaftliche Buchgesellschaft, Darmstadt.

Church, A.

1935 An unsolvable problem of elementary number theory. Preliminary report (abstract); Bulletin of the American Mathematical Society 41, 332–333.

1936 An unsolvable problem of elementary number theory; American Journal of Mathematics 58, 345–363.

Dedekind, R.

1854 Über die Einführung neuer Funktionen in der Mathematik (Habilitationsvorlesung); translated in: *From Kant to Hilbert*, volume 2, (W.B. Ewald, ed.), Clarendon Press, Oxford 1996, 754–762.

1872 *Stetigkeit und irrationale Zahlen*; Vieweg, Braunschweig; translated in: *From Kant to Hilbert*, volume 2, (W.B. Ewald, ed.), Clarendon Press, Oxford 1996, 765–779.

1888 *Was sind und was sollen die Zahlen?*; Vieweg, Braunschweig; translated in: *From Kant to Hilbert*, volume 2, (W.B. Ewald, ed.), Clarendon Press, Oxford 1996, 787–833.

1890 Brief an Keferstein; translated in: *From Frege to Gödel* (J. van Heijenoort, ed.), Harvard University Press, Cambridge, Massachusetts, 1967, 98–103.

Gandy, R.

1980 Church's Thesis and principles for mechanisms; in: *The Kleene Symposium* (J. Barwise, H.J. Keisler, and K. Kunen, eds.), North-Holland Publishing Company, Amsterdam, 123–148.

Gödel, K.

1931 Über formal unentscheidbare Sätze der Principia Mathematica und verwandter Systeme I, Monatshefte für Mathematik und Physik 38, 173-198; reprinted and translated in: *Gödel – Collected Works*, volume I, Oxford University Press, 1986, 144–195.

1946 Remarks before the Princeton bicentennial conference on problems in mathematics; in: *Gödel – Collected Works*, volume II, Oxford University Press, 1990, 150–153.

Hilbert, D.

1900 Über den Zahlbegriff; Jahresbericht der Deutschen Mathematiker Vereinigung 8, 180–194.

1905 Über die Grundlagen der Logik und der Arithmetik; translated in: *From Frege to Gödel* (J. van Heijenoort, ed.), Harvard University Press, Cambridge, Massachusetts, 1967, 129–138.

1918 Prinzipien der Mathematik; Lectures from WS 1917/18; unpublished notes written by Paul Bernays.

1922 Grundlagen der Mathematik; Lectures from WS 1921/22; unpublished notes written by Paul Bernays.

1923 Die logischen Grundlagen der Mathematik; Mathematische Annalen 88, 151–165.

Hilbert, D. and Bernays, P.

1934 *Grundlagen der Mathematik I*; Springer Verlag, Berlin.

1939 *Grundlagen der Mathematik II*; Springer Verlag, Berlin.

Sieg, W.

2002 Beyond Hilbert's Reach?; in: *Reading Natural Philosophy* (D.B. Malament, ed.), Open Court, Chicago, 363–405.

2007 Church without dogma: axioms for computability; to appear in: *New Computational Paradigms*, B. Loewe, A. Sorbi, and B. Cooper (eds.), Springer Verlag.

Sieg, W. and Field, C.

2005 Automated search for Gödel's proofs; Annals of Pure and Applied Logic 133, 319–338.

Stein, H.

1988 Logos, Logic, Logistiké: Some philosophical remarks on the 19th century transformation of mathematics; in: *History and Philosophy of Modern Mathematics* (W. Aspray & P. Kitcher, eds.), Minneapolis, 238–259.

Turing, A.M.

1936 On computable numbers, with an application to the Entscheidungsproblem; Proceedings of the London Mathematical Society, series 2, 42, 230–265.

Zermelo, E.

1908 Untersuchungen über die Grundlagen der Mengenlehre I; translated in: *From Frege to Gödel* (J. van Heijenoort, ed.), Harvard University Press, Cambridge, Massachusetts, 1967, 199–215.

23
William Tait
Professor Emeritus

University of Chicago, USA

A Road to Philosophy and Mathematics

I became interested in philosophy and mathematics at more or less the same time, rather late in high school; and my interest in the former certainly influenced my attitude towards the latter, leading me to ask what mathematics is really about at a fairly early stage. I don't really remember how it was that I got interested in either subject. A very good math teacher came to my school when I was in 9th grade and I got caught up in his course on solid geometry; but he soon left and math then lost its luster again in the hands of teachers who neither liked nor understood it. Calculus wasn't taught in high school in those days, or at least not in mine: besides geometry we learned some algebra (how to solve some equations) and trigonometry (with, of course, very little proved). I doubt that even the *word* "philosophy" passed the lips of any of my teachers. My mother, who worked for a publishing house, brought home for me copies of, among other works, the Jowett translations of Plato's *Dialogues*, Will Durant's *Story of Philosophy* and Courant and Robbins' *What Is Mathematics?*; but I can't remember why she did that: She wasn't at all intellectual and, as far as I recall, my interests at the time were mostly confined to sports and girls—in some order. Maybe she just thought it was time for me to develop new interests.

After high school, I went in 1948 to Lehigh University, then at least primarily an engineering school, on an athletic scholarship (which I was lucky to get: I wasn't that good an athlete and there was a glut of more talented GI's returning to school). There I had the good fortune in my first year to have an introduction to philosophy course with Lewis White Beck. He had just moved there from the University of Delaware and shortly thereafter moved on to the University of Rochester, where he became one of the leading lights of American Kant studies. My good luck was compounded when, in my second year, Adolph Grünbaum arrived at Lehigh, fresh from graduate school at Yale, and stayed at least long enough for me to graduate, before moving to the University of Pittsburgh as Andrew Mellon Professor of Philosophy of Science. Beyond his excellent course on the philosophy of science, in which I was exposed to the main currents of thought on the subject at that time, he was both an important source of encouragement and of enormous help to me in getting started as a scholar: He introduced me to Theodore Hailperin, who taught me some logic (I can't remember whether it was in a regular course or in a special reading course) and to another young member of the mathematics department, Samuel Goldberg, who met with me every week to study, eventually, Hardy's *Pure Mathematics*—a wonderful, although humbling, experience. I should also mention that Adolph, along with other members of the Philosophy department at Lehigh then, were extraordinarily supportive and helpful in enabling me to transfer from an athletic scholarship to an academic one: With an injury in the spring of my sophomore year and, probably more importantly, a radical change in my interests, going out every afternoon to get pounded had lost a lot of its appeal.

Following Adolph's example, if not his advice (I can't remember), I went to Yale as a graduate student in philosophy in 1952. In my first year, I took in sequence two semester-long courses in set theory. In the first of them, Fredrick Fitch began to formally develop the Gödel monograph *The Consistency of the Axiom of Choice and the Generalized Continuum Hypothesis with the Axioms of Set Theory* in his system of natural deduction. I can't remember how far we got; but given the fact that the details in that monograph already threaten to overwhelm the ideas, the addition of formal deductions did little to lend light. Following that, John Myhill gave a course in foundations of set theory in which he compared various axiomatizations—*ZF*, The predicative second-order systems of von Neumann and Bernays/Gödel, Quine's *New Foundations*, etc. The course slightly swamped me and when, after I had (more or less) finished a difficult final exam, Myhill asked me to lend him a quarter to buy a beer, I almost strangled him. (Incidentally, you got quite a big glass of beer for a quarter in those days.) John went on leave the next year and never returned to Yale—a large-size loss for me. I had a Fullbright fellowship in my third year to go to Amsterdam to study intuitionism with, I thought, Brouwer; but he had retired by the time I arrived and my study of intuitionism was confined to some lectures by Heyting and talks with his assistants and students. The most interesting activity in logic was a short series of lectures by Leon Henkin, also there on a Fullbright as I remember, on cylindrical algebras. Nevertheless, I had a profitable time in Amsterdam, beginning a serious study of mathematics, which I continued when I returned to Yale. Logic in New Haven at that time was represented by Fitch, whose main interest was in a variation of combinatorial logic, which he called *basic logic*, and Alan Ross Anderson, who had been a fellow student in Myhill's class that first year but who returned after a few years as an assistant professor. I very much admired Alan, but his interests were then primarily in modal logic and so remote, or so I believed, from foundational issues. (In spite of suggestions about introducing modality into mathematics, I still believe that.) I was working my way through Kleene's *Introduction to Metamathematics*, but entirely in isolation: I remember hours of confusion because I failed to recognize that, e.g., of two upper case "A" 's involved in the same argument, one was italicized and the other not. (It is not a very humane book.) I also remember thinking for most of a week that I had gone mad, because due to a misunderstanding of the statement of Gentzen's *Hauptsatz*, I thought I had a very elementary proof of the consistency of first-order arithmetic. The philosophy department itself was at that time, I felt, in serious decline.[1] My discussions with Fitch about combinatorial logic/lambda calculus (we never talked about my work) probably served me well: it became a staple for me in thinking about various problems in proof theory. But on the whole, although I had many very bright fellow-students, I found that the interests represented by the other members of the philosophical community there were generally quite remote from mine.

I believe (but am not certain) that I remained in logic/philosophy and went on to obtain a PhD in that field only because of the Summer School in Logic at Cornell in 1957, the first I believe of its kind. Although philosophy remained (and remains) my main interest, my experience in graduate school did not lead me to high expectations for life in a philosophy department, and I had been drifting away from the subject. I stayed at Cornell for the

---

[1] For some people, the decline began somewhat later, in the 1960's: It probably depends upon what their interests in philosophy were. There is, however, general agreement that there was a serious and long-lasting period of decline beginning at least in the 1960's, but also that the process reversed and that the present department is quite strong.

first five of the six-week program and left with my head spinning. At that time, journals in logic were years behind the frontiers of the subject and, after the isolation of Yale, I had had no idea of the riches I began to glimpse. I spent quite a bit of time, as I remember, speaking with Anil Nerode, who wasn't that much older than I, but light-years ahead of me in matters logical. He was very encouraging and it was he who convinced me that some things I had worked on, computable second-order functions and restricted forms of Turing reducibility for them, might actually be of interest. (I never tested the conviction, however: It all went into my dissertation, which I wrote in the summer of 1958 and, after defending it the following autumn at Yale, never looked at again.) I also remember evenings listening to Paul Halmos and Alfred Tarski, to both of whom I have remained grateful for the time that they spent with students at Cornell that summer. It was there, too, that I was exposed, primarily through Georg Kreisel's lecture on Gödel's *Dialectica* interpretation, to the possibility that there still remained after Gödel's incompleteness theorems a program of constructive interpretation of classical mathematics—the possibility that my taste for logic could be comfortably united with my feeling that philosophy is, after all, the serious matter.

It was pure coincidence that, a year later, I connected up with Kreisel at Stanford. My interest in logic had been rekindled and back at Yale I was working my way through volume 2 of Hilbert and Bernay's *Grundlagen der Mathematik*. The job offer from Stanford, probably engineered by Alan Ross Anderson, was the second one I had in the academic year 1957–8 and, as I did the first, I was inclined to turn it down.[2] I had never heard of Stanford (and so of course had no idea that Kreisel was about to begin a part-time appointment there) and, although graduate students of today will find this hard to believe, life as a graduate student in those days was very pleasant: There was almost no tuition and, with a little bit of teaching, one could live quite comfortably, studying the things one wanted to study, without the hassle of a real job. But ARA began to get seriously angry with me – as did quite possibly my wife as well – and so off we went. I have certainly never regretted it: Besides Kreisel, Sol Feferman, whom

---

[2] In those days, there wasn't much formality – or evenhandedness – about hiring: If a department wanted to recruit, someone simply called up his favorite department and asked who they had available. I had never previously met any of my new colleagues at Stanford in philosophy nor had they, I believe, previously ever read a word that I had written.

I had met at Cornell, had arrived at Stanford the previous year; and through our logic seminar and what was then a very close connection with the logic group at Berkeley, we had through the years I was there (up to the summer of 1965) a rich assortment of logicians hanging about at any time.

But the program of constructive foundations of classical mathematics did not in the end fare so well. Spector's extension of the *Dialectica* interpretation to second-order number theory using bar recursion of higher types (1961) and Takeuti's consistency proof for Π_1^1 analysis (appearing in unpublished form around 1964) were the highpoints. But one could find no grounds for accepting higher order bar recursion as constructive and Takeuti's proof – essentially a proof that cuts can be eliminated from deductions in Π_1^1-CA with the ω-rule – proceeds by showing that a certain quite unintuitive system of ordinal notations is well-founded, the proof of which can in no reasonable sense be termed constructive. My own program of attempting to constructively interpret second-order number theory using the epsilon-substitution method bit the dust in the winter of 1962–3. (In my defense, I wasn't the only one naive enough to think that such a result was obtainable: There wasn't then the same clear sense we have now of the limits of constructive methods as we then understood them.) I was at IAS in Princeton at the time and one fallout of my discussions with Gödel about my failure was his suggestion to me that one should consider what instances of second-order comprehension could be satisfied in a theory of inductively defined sets. I don't know whether this was the source of the initiation of studies of iterated inductive definitions at Stanford around that time; but it was for me. For it was immediately clear that the classical theory of finite iterations of inductive definitions of sets of numbers was sufficient to satisfy $\Pi_1^1 - CA$ and almost as immediately clear that a partial cut-elimination result for that theory with the ω-rule – the elimination of cuts in deductions of purely arithmetic formulas – was provable in the constructive version of that theory.[3] In other words, it was possible to end-run around Takeuti's argument: I doubt that I was the only one to sigh in relief that one didn't have to learn that awful argument. But alas, as Harvey Friedman

---

[3] My notes on this are in the form of copies of letters to Kreisel, dated in 1966. In fact, I had noted that, further iterating inductive definitions, one could embed the *rule* of Δ_2^1 comprehension. I lectured on these things at Rockefeller University in February 1967 and discussed it fairly broadly through the summer of 1967.

pointed out to me at the *Buffalo Conference on Intuitionism and Proof Theory* in 1968, $\Pi_2^1 - CA$ is a barrier for iterated inductive definitions. The least ordinal α for which the second-order version of L_α satisfies $\Pi_2^1 - CA$ is non-projectible.

For me, after thrashing around for a year or so, that was (until recent times) the end of proof theory: It seemed impossible that constructivity as we understood it had the resources for interpreting classical second-order number theory. In the light of work on Martin-Löf's type theory and intuitionistic set theory, that judgment might have been premature. It is also the case that proof theory survived as a purely mathematical theory. For example, the techniques of proof theory may be used to extract information implicit in classical proofs—an application of proof theory, called *proof-mining*, that was initiated by Kreisel in the early 1950's and has been pursued in recent times by Kohlenbach and others. It is also the case that Rathjen has recently extended Takeuti's result by obtaining the proof-theoretic ordinal of $\Pi_2^1 - CA$. But it still remains to be shown that proof theory has any remaining and redeeming philosophical virtue.

Of course, constructive mathematics itself flourished: Perhaps the most important development in that field was the publication of Errett Bishop's *Foundations of Constructive Analysis* in 1967, although good work has continued to be done in Brouwer-style analysis. One reaction of people who had worked in the program of constructive foundations for classical mathematics to its failure was to feel "So much for classical mathematics!" and to thenceforth restrict their attention to something less than the latter. I did not share that reaction: I appreciate the attraction of constructive mathematics, but the game was (is) to understand classical mathematics.

Abandoning proof theory was defining myself as primarily a philosopher. Even if one felt driven by philosophical motives, Hilbert's program, even in its extended form, gave one something to do without philosophical reflection: Reduce mathematics based on the axiomatic conception to mathematics based, if not on the conception of Kant, Kronecker and Hilbert/Bernays, nevertheless on a reasonable extension of those ideas (allowing non-algorithmic properties, but, if one proved the existence of an object with such a property, one could extract an algorith for it). Its failure left for me an itch, but I didn't know where to scratch. I wanted to try to understand why we felt that something needed to be said or done about the foundations of mathematics, and what it really was that needed to be done. So I began, in the mid-1970's, a study in philosophy of mathematics and, equally importantly, in the history of the development of the central concepts of mathematics: number, function, and set.

The Role of Mathematics in Philosophy

Aside from the negative business of letting the fly out of the fly-bottle, I think of philosophy as being primarily foundational: it has no subject matter of its own, but rather refers to a characteristic way of approaching the sciences—physical, biological, social, cognitive, and mathematical—like Plato's dialectician, seeking clarity and the first principles of each science. Where works in philosophy appear to be advancing theories about language, the mind, or whatever, I tend to see nascent science at best and bootleg science at worst.[4]

However, I don't want to entirely downplay the freeing of flies, nor do I think that it is entirely divorced from philosophy in its foundational role. For example, in the foundations of mathematics itself, historical resistance to the actual infinite was based upon supposed paradoxes, including in recent times the so-called 'paradoxes of set theory', which have all been seen to be based upon confusion. Yet the resistance persists on other grounds. Thus, the pursuit of the actual infinite leads us to speak of the existence of objects which are not simply in themselves infinite, but also cannot even be effectively approximated by finite things. There are those who not simply choose to pursue constructive mathematics, which avoids such objects, but argue that speaking of them is meaningless or wrong. There are others, in this case, perhaps, more often philosophers quite removed from mathematics itself, who advocate some form of nominalism, on the grounds that, no matter how internally coherent mathematics might be, it speaks of 'abstract objects', and these simply don't exist. In both cases, I believe, there are captive flies buzzing around. Most of my own non-technical publications have been of the freeing-of-the-flies variety, perhaps the most notable example being "Proof and truth: the 'Platonism' of mathematics". But what I want to point out here is that this enterprise is not totally unconnected to the foundational enterprise. Many of those who have been involved in the

---

[4] Of course I am excluding here the many instances of philosophers who publish work that they quite frankly see as having scientific content in the usual sense, subject to the usual critical standard.

development of set theory, itself, for example, have been afflicted with 'philosophical' doubts about the existence of sets—Tarski's well-known 'finitism' being a case in point. Freeing the *internal* problems of foundations from these external and groundless concerns is surely part of the foundational role of philosophy.

I do not see that mathematics itself plays any role in this more therapeutic kind of philosophy. So the question of the use of mathematics in philosophy can only be, for me, the question of its use in foundations of science. The idea of 'mathematizing' a science, of creating mathematical theories which idealize the phenomena in question—that is, in terms of which we can understand and reason about these phenomena and control them, goes back at least as far as Plato. In near contemporary times its great exponent was Hilbert, and the ubiquity of titles in the last century and earlier of the form "The mathematical foundations of ..." attests to the success of this approach to foundations of science.

Because of a primarily foundationalist conception of philosophy, I include in the field much more (and much less) than the twentieth century division of the disciplines allows for. In particular, many works on foundations of mathematics, that are generally counted only as mathematics—*unless they are very old*—I think of as philosophy and hence as instances of the 'use of mathematics for philosophy'. A few examples: Riemann's "On the hypotheses which underlie geometry", Dedekind's "Continuity and irrational numbers" and "The nature and significance of numbers", Hilbert's *Foundations of Geometry*, Cantor's theory of transfinite numbers in *Foundations of a General Theory of Manifolds: A Mathmatico-Philosophical Investigation into the Theory of the Infinite*, Frege's analysis of quantification in his *Begriffsschrift*, the whole nineteenth century movement in the foundations of function theory and search for the proper definition of the integral, culminating in Lebegue's "On a generalization of the definite integral", the works of Zermelo, von Neumann and Gödel on foundations of set theory—as well as much of the contemporary work in this area. The analysis of computability in the works of Turing and others belongs in this list. And of course, if we move to the foundations of other sciences—physics, biology, economics, etc., a whole class of other examples come to mind.

Of my own work, perhaps the paper "Finitism" may be regarded as an application of mathematics to philosophy, in that it attempts to give an analysis of a particular conception of mathematics in terms of the formal system of primitive recursive arithmetic. Indeed, I begin to believe that, independently of whatever version of Kantianism Hilbert and Bernays were drawing on in their conception of finitism, primitive recursive arithmetic is the genuine heir of Kant's conception of mathematics—indeed, of a pre-nineteenth century constructivist conception to which Kant gave voice. (My view of Kant in this respect is heavily influenced by Michael Friedman's work on Kant's philosophy of mathematics.) My work in proof theory in the 1960's and early 1970's was in aid of a philosophic program; but, whatever intrinsic value that work has, the philosophic program failed. Moreover, the program presupposed a radical difference between constructive mathematics and classical mathematics, the former based on an idea of construction, the latter based upon an idealized domain which we access by axiomatically describing it. As a result of subsequent philosophical reflection (of the freeing-of-the-flies variety), I no longer believe that: I don't see constructive mathematics as based on a different conception of mathematics but as, basically, a subdomain of classical mathematics. I first discussed this in a (badly written) paper "Against intuitionism: Constructive mathematics is part of classical mathematics." A further discussion is in the introduction to a collection of my philosophical essays, *The Provenance of Pure Reason: Essays in the Philosophy of Mathematics and its History.*

The Proper Role of Philosophy of Mathematics in Relation to Logic, Foundations of Mathematics, Mathematics, and Science

In its positive, foundationalist, guise, the proper role of philosophy and, specifically, philosophy of mathematics in foundations of mathematics, mathematics and science is obvious. Much of the work in philosophy of mathematics of the last century was concerned with foundations in this sense; and indeed, it continues to this day.

Some of it, in the logicist program and Hilbert's program (and its extension) in particular, was concerned to give a foundation for all of classical mathematics. The logicist program in the sense of Frege, of reducing mathematics to logic, was of course doomed to failure. In the sense of Dedekind, however, in which the aim was to eliminate intuition and replace it by logical analysis, it led to the modern conception of mathematics as based on the axiomatic method. This in fact set the stage for Hilbert's program: To prove the axioms consistent or, in the case of the extended program, to interpret the theorems of classical mathematics as theorems of constructive mathematics. The failure of this program was not so immediate. It should be mentioned, too, that its failure is not the failure of the axiomatic conception of mathematics—and that is fortunate, since it is the only viable conception we have. Rather, the significance of Gödel's second incompleteness theorem is that it is a fact of mathematical life that we are forever at risk of encountering a contradiction.

Incompleteness, on the other hand, is the engine driving contemporary foundations of classical mathematics, i.e. philosophy of mathematics in the positive sense, although the work done in this area is quite technical and is not philosophy as the term is usually used: There are many questions that the axioms of *ZFC* do not suffice to settle and so one is led to believe that the axioms do not sufficiently express the conception of a universe of sets obtained by iterating the powerset operation and, following the suggestion of Gödel, one would like to find axioms expressing even higher iterations of this operation that will lead to the solution of open problems in everyday mathematics. The discovery that certain of these large cardinal axioms yield the solution of problems in descriptive set theory, such as whether all projective sets are Lebesgue measurable and whether they have the property of Baire, has been one great success in this direction and leads to the hope that further axioms expressing even higher iterations of powerset will lead to solutions of, say, the Continuum Problem.

But there were and are, too, revisionary programs aimed at restricting the scope of mathematical reasoning. One example is the predicativism developed by Weyl and, later, Feferman. Another is the strict finitism of Kronecker, in which the objects of mathematics are restricted to those representable by whole numbers and whose concepts are restricted to those equipped with algorithms for determining which objects fall under them—the position that Hilbert adopted as the methodological stance upon which to prove the consistency of axiomatic mathematics, and the more liberal constructivism of Brouwer, of Weyl, and, later, of Bishop. Of course, both predicative and constructive mathematics can be pursued as interesting domains of investigation in their own right—subdomains of classical mathematics; but I am referring here to a stance according to which we *ought* to adopt a more restrictive kind of mathematics. The arguments for this have been various: For example, in the early part of the last century, such as in the writings of Weyl, they were often based on the so-called 'paradoxes of set theory'. Brouwer also referred to these 'paradoxes' in his polemic against classical mathematics; but his more positive argument (and one would suppose this to be so of Kronecker, too, if he had chosen to write more on the subject) appealed to an earlier tradition in which, at least if one sufficiently hid epsilon-delta arguments behind infinitesimals, one could believe in the picture of mathematics presented by Kant, that all of mathematics consists essentially of construction according to rules. In more recent times there has been Michael Dummett's argument for constructive mathematics based upon a theory of meaning. Also, in philosophy of mathematics itself, largely in isolation from the actual practice of mathematics, general and *a priori* views on ontology—about the existence of what some writers call 'abstract objects'—have led to the charge that mathematics or at least some parts of it are meaningless or false and/or to the view that at least a part of it can be understood only as a formalism.

My own non-technical papers in philosophy of mathematics, other than some of them of primarily historical content, are philosophical in the negative sense: Their primary aim has been to disarm the arguments behind these revisionary programs. In one direction, I have attempted to counter the idea that constructive mathematics is a different subject from classical mathematics and have argued that one can understand constructive mathematics as a subdomain of classical mathematics. In several papers and in my collection of essays *The Provenance of Pure Reason: Essays in the Philosophy of Mathematics and Its History*, I pointed out that the conception of meaning that Dummett believes to support intuitionistic mathematics is equally compatible with the classical conception and that the apparent constructive refutations of classical theorems often referred to are in fact simply a matter of changing the meanings of words—that with a certain disambiguation, the 'counterexamples' are classically valid, too. In another direction, I have attempted to show that the qualms about the existence of 'abstract objects' that have led some philosophers and mathematicians to reject or at least to question parts of mathematics are based upon an illusion that there is some univocal notion of existence on the grounds of which we can legitimately argue for or against the existence of mathematical (or physical or mental) objects. This is a lesson that I learned from Wittgenstein's *Philosophical Investigations*, although, paradoxically, his own views about mathematics were so out of sync with the actual mathematics of his time that he failed to apply his own lesson. Finally, by arguing, again following a line of thought I believe to be in Wittgenstein's *Investigations*, that Hilbert's conception that the (categorical) axioms define the mathematical structure – that the objects of the structure are, so to speak, constituted in the axiom system – is not entirely different from the sense in which the objects of our daily life are constituted in the language in which we speak and think about them, I have attempted to disarm the charge of formalism that has been leveled against the axiomatic conception of mathematics.

Late 20th Century Philosophy of Mathematics

It is disappointing to me, now in the twenty-first century, that so many of the ghosts that haunted philosophical discussions of mathematics at the beginning of the twentieth century are still with us. At the beginning of that century, the concepts of set and (in our sense) function and, generally, the explicit acceptance of the actual infinite (in the sense, not of there being infinitely many things—a potential infinity, but of there being infinite things) were still relatively new in mathematics: A new language had to be learned and old misconceptions and fallacious 'paradoxes' had to be exposed. True, the latter had already been done in Bolzano's *Paradoxes of the Infinite* and in Cantor's *Foundations of a General Theory of Manifolds: A Mathmatico-Philosophical Investigation into the Theory of the Infinite*, but the latter of these was of relatively recent vintage (1883) and apparently not much read by philosophers and Bolzano's work, because of, ultimately, an inadequate notion of a set, failed to lay to rest the ancient 'paradox' concerning 'unequal infinities', i.e. sets of the same size as one of their proper subsets, and well as those problems that arose from failing to distinguish between what we would call structures and their underlying sets.

Of course, the appearance of the *new* 'paradoxes of set theory' contributed to the sense that the new language might turn out to be incoherent; but a conception of set theory having as its models a potential infinity of universes of sets (where each universe appears as a set in another one and there are no absolute 'proper classes') and which is not in the least subject to these paradoxes has been in existence since Zermelo's 1930 paper.

Resistance to accepting the new language has alas been reinforced by superstition concerning the issue of "what there is," where this is taken to be, not an issue *internal* to the language or theory in which the objects are purported to make their appearance, but an external question concerning the legitimacy of the theory itself. In United States and England, at least, the hegemony of W.V. Quine among philosophers on the subject of mathematics through much of the last half of the century had a lot to do with this resistance. I'm referring here not only to his unwarranted "common sense is bankrupt" point of view concerning set theory, but also his views about ontology. His slogan, "To be is to be the value of a bound variable," turned out to be a *criterion* for ontological commitment of theory, one which mathematics might fail to satisfy, rather than a banishment of the issue of ontological commitment to mathematical objects (sets, functions, numbers, etc.) from consideration entirely, as it might and should have been. The misfortune was compounded by Quine's view of the role of mathematics in natural science. As opposed to the view that Euclidean geometry, arithmetic, and set theory concern their own ideal domains, Euclidean space, the system of natural numbers, and suitable universes of sets, respectively, *independently of any possible applications that they might have in our theories about the natural world*, Quine took the position that mathematics has no autonomous status and that its validity rests holistically with its role in natural science. This view, too, framed many of the topics of discussion in the last part of the century, which were, therefore, far from any involvement with real issues concerning mathematics.

The same lack of involvement may be ascribed to the contemporary neo-logicism. When it is considered, not as a possibly interesting—though surely quite limited—investigation in its own right, but as an alternative to mathematics as it is being practiced, one is moved to ask: Why? The motivation for it as a better alternative for doing arithmetic and analysis seems based upon the same monochromatic conception of existence as Quine's. (In fact, it goes back through Frege to Kant and ultimately to Aristotle, and is opposed to the tradition, going back through Leibniz to Plato and forward through Dedekind and Cantor to Hilbert, according to which mathematics concerns ideal domains.) Frege, realizing that the demands of mathematics in his time required that whole numbers be regarded as objects, needed to make a correction in Kant's philosophy, one that would admit numbers into the *same* universe that Kant had wanted to restrict to things representable in sensible intuition. The neo-logicists seem committed to the same view: whole numbers and real numbers, say, are part of the same universe as physical objects, arising out of equivalence classes of concepts that are meaningful for all objects—so that, for them too, it makes sense to ask whether Julius Caesar is a number! The difference being that, instead of Frege's inconsistent assumption that arbitrary extensions of concepts belong to the universe, they make the more modest assumption that this is so (essentially) of suitable equivalence classes. As a philosophical stance, it seems sterile; as for the development of the theory itself, it lacks the kind of connection with actual mathematics that constructive mathematics has, for example, as a style of proof in which existence proofs yield algorithms.

In speaking about philosophy of mathematics in the late twentieth century, one certainly needs to mention the influence of Gödel. Aside from his technical work, his 1948 paper "What is Cantor's continuum problem" along with the supplement of 1964 have been quite influential both in foundational work in set theory and, alas, in muddying the waters over the issue of 'what there is' with his subscriptions to 'Platonism'. In the former respect, I have already mentioned that his view that the pursuit of large cardinal axioms might lead to the solution of mathematical problems has served as motivation for research in set theory and, indeed, has born fruit—although not with respect to the problem at issue in that paper. The publication of his collected works has led to fairly intense discussion of his philosophical views in recent times; and having contributed rather more substantially to that discussion than I ever intended, perhaps I can beg off discussing it further here. One matter though that I would like to mention is his interest in Husserl's phenomenology, which he began to study, it seems, in the late 1950's. I don't know how much more there is to be found out about it in the Gödel archives; but it has attracted considerable attention among phenomenologists and it will be interesting to see what might develop from it.

The Most Important Open Problems in the Philosophy of Mathematics and the Prospects for Progress?

For me, the most important open problem in philosophy of mathematics is in foundations of mathematics, and that is the search for new axioms of set theory—which means, too, the search for grounds for accepting them. There are many interesting directions of development in logic, but in philosophy of mathematics, I believe that this is the overwhelmingly most important problem. But it is a problem now largely in the hands of set theorists. Maybe one important open problem for those of us who are primarily philosophers is that of gaining access to that problem.

24
Albert Visser

Professor of Philosophy

University of Utrecht, The Netherlands

Why were you initially drawn to the foundations of mathematics and/or the philosophy of mathematics?

The story starts somewhere between 1969 and 1974. I was studying applied mathematics at the University of Twente, a technical university in the east of Holland. I was drawn to philosophy for three reasons.

The first reason was the process of rebelling against my protestant-religious upbringing. My parents were protestants of an enlightened sort. They did not believe that the snake did speak. On the other hand they were very serious about their faith. The school I attended, het Johannes Calvijn Lyceum, was of a more dogmatic persuasion. For many teachers that snake did indeed speak. When I gave a presentation on the theory of Evolution, the teacher refused to give me a mark, because she thought that giving a high mark would seem like recognition of the ideas presented. Giving a low mark, on the other hand, would not do justice to the quality of the presentation. (In hindsight I am struck by the honesty and integrity of her choice.) During the first year of my study at the Technical University of Twente, I became an atheist. I embraced a positivist philosophy which provided the intellectual tools to criticize the faith of my youth. My father placed great value on argument, so he and I had endless discussions on religion. These discussions increased my interest in philosophy.

Secondly, the technical university was not quite the right place for my theoretical inclinations. So I compensated by studying some more theoretical things, like foundations of analysis (a wonderful course by W.W.E. Wetterling) and courses in philosophy.

Thirdly, in the context of this particular technical university at this particular time, both mathematics and philosophy were *progessive*, where e.g. mechanical engineering was *conservative*. I naturally sided with the progressive group. We were members of Amnesty and protested against the exploitation of the environment.

While still in Twente, I studied philosophy under Errit van der Velde and Louk Fleischhacker. I was most influenced by Louk Fleischhacker. Louk was a student of both Haskell Curry and Jan Hollak[1], truly an intriguing combination of teachers. Louk's subject was philosophy of mathematics. In his classes he simply sat in front of the students and started to philosophize. I was deeply impressed. In hindsight I cannot have understood much of what Louk was saying. Moreover, I was much more drawn to the more analytical style of philosophy. Still, I find that things Louk said pop up whenever I am thinking about philosophical problems.

The first philosophical book I read was Russell's *Enquiry into Meaning and Truth*, the second Popper's *Logic of Scientific Discovery*, the third Quine's *Word and Object*. For my further career Quine's was the most important influence. Of course, first I was a Quinean, later an anti-Quinean. Today, I still feel that Quine's is an absurd philosophy, but I do think he is a great philosopher.

Towards the end of my studies in Twente, I had realized that both my inclination and my talent where in Logic and Philosophy. So after, in 1974, I obtained the degree of Bachelor in Applied Mathematics, I moved to Utrecht to study those subjects. My teachers there were Dirk van Dalen, Henk Barendregt, and, for some time, Craig Smoryński and Jeffrey Zucker. But now the story moves beyond early beginnings and I stop here.

What examples from your work illustrate the use of mathematics for philosophy?

Around 1980, I read Kripke's *Outline of a Theory of Truth* ([Kri75]). The paper was a revelation to me. Here we found a fascinating interplay between philosophical considerations and interesting metamathematical methods. From the technical point of view, the paper provides a nice way to represent inductive sets over the given acceptable structure. Moreover, Kripke's paper highlights the interest of the other fixed points. Kripke introduces the maximal intrinsic fixed point, which was independently invented by Manna

---

[1] Jan Hollak was a Dutch philosopher whose philosophy can be, very roughly, characterized as neo-Hegelian.

and Shamir, see [MS76]. The idea of intrinsicity yields a Zorn-free proof of the existence of fixed points of monotonic functions on a ccpo. From a philosophical point of view, the paper offers various criticisms of the Tarski approach, the ideas of groundedness and intrinsicity, and the insistence that a solution should be systematic and global in the sense that it should cover a wide range of self-referential phenomena. This last point illustrates one advantage of a logico-technical approach: we have a precise specification of the range of cases that we are treating.

An interesting aspect of Kripke's paper is his attention to how we actually think. This attention presents a break with the tradition of Frege, Russell, Quine and Tarski, who rather wanted to improve on ordinary language for certain special purposes.

My enthusiasm resulted in a paper [Vis84], and a Handbook article [Vis89] (or, in a slightly corrected version: [Vis04]).

In the light of strengthened versions of the Liar, it becomes evident that Kripke's paper does not offer us a solution in any final sense. Moreover, I became convinced that the fundamental form of the paradox is the propositional one. Sentences are true or false in a derived way: they express propositions that are true or false. Kripke argues as follows. If truth is primarily sentential, it is clear that there is a sentential problem of the Liar. If truth is primarily propositional, then there is still the sentential Liar: *this sentence expresses a false proposition*. So, in all cases, it is sensible to study the sentential Liar. The loophole here is that the structure *a sentence expresses a proposition which has a truthvalue* imposes constraints on solutions. Moreover, I guess, the problem of truth *is* the problem of the proposition. To articulate what a proposition is, ipso facto, to explain what it is for it to be true.[2] Jon Barwise and John Etchemendy tried to give a solution of the Liar in its propositional manifestation. See [BE87]. Regrettably, the explication of the notion of proposition in [BE87] is not all that convincing. Also the framework of Barwise and Etchemendy is underspecified in the sense that it does not give us precise instructions on how to analyze concrete examples. Still it is a sympathetic attempt.

I will not deny that there has been a lot of good work on the Liar since Kripke; nor has this work been devoid of applications —e.g. to negation in Logic Programming and to transfinite Turing

---

[2] I would resist the converse, which would in effect identify the proposition with its truth conditions.

machine computations. Still I feel that the harvest of insight on the Liar front from 1975 on has been disappointing. There is still an ununderstood kernel that all this work does not come close to touching.

A paper that deeply influenced me is Groenendijk & Stokhof's [GS91]. In this paper, the authors develop a version of predicate logic, called *Dynamic Predicate Logic* or *DPL*, that has scope conventions that differ from those of ordinary Predicate Logic. Specifically, the scope of the existential quantifier is not constrained to the formula of which it is the main connective. The basic semantic idea is that the compositional meanings of our formulas are not sets of assignments (constraints on a space of items) but reset relations (things that model semantical actions, context change potentials).

I was struck by the simplicity and elegance of DPL. I felt that it clarified ordinary Predicate Logic by making the specific choices behind its scope conventions visible. Also, for DPL, the relationship between Propositional Logic and Predicate Logic is clearer: Dynamic Predicate Logic becomes just a specific theory *in* Dynamic Propositional Logic.[3]

Some of the conventional wisdom of ordinary Predicate Logic fails for DPL. For example in Predicate Logic we have two kinds of occurrences of variables: free ones and bound ones. In DPL we have at least three kinds of occurrences. I tried to make this explicit in my paper [Vis89]. Thus, DPL touches on the question of the nature of the variable.

A step that Groenendijk and Stokhof never made was to draw the consequences of their approach for syntax. There are two reasons for this. First the Amsterdam School conceives of itself as occupied with *semantics*. Secondly, Groenendijk and Stokhof sought continuity with predicate logic on the surface level. They – I think erroneously – tried to make DPL look as much as possible like ordinary Predicate Logic. (I think that the connection lies at a deeper level: that of a fibered category of Boolean algebras.) I felt that DPL strongly suggests a different view of syntax: flat, as incremental as possible, with a blurring of the boundary of the categorematic and the syncategorematic. I wrote some papers on

---

[3] I think that the algebraic approach connected to DPL has some definite advantages over Cylindric Algebras. Marco Hollenberg and I did some work on these algebras. See [Hol97], [Vis97] and [HV99]. Regrettably it never caught on.

the subject but regrettably always failed to give the leading idea the simple directly understandable form that it seems to have before my mind's eye. See e.g. [VV96], [Vis01], [Vis02] and [Vis03].

If one asks: specifically for the Philosophy of Mathematics, *what is the contribution of DPL?* I would answer as follows. (i) It is one way of bringing out specific choices made in the design of ordinary predicate logic. As such it may help us not to draw consequences which rest on these specific choices too easily.[4] A good example of the freedom created by the dynamic approach is the treatment of definite descriptions. In a dynamic approach we can treat descriptions as constrained anaphors. Moreover, we can escape from the necessity to make the constraint on the values purely contentual: we can bring in salience as a property of a referent in a discourse. (ii) It brings out the importance of taking a new look at the theory of syntax and the question of the nature of the variable.

Let me finally mention the philosophical question *what is number?* Frege came up with an answer that I always found strongly compelling.[5] His idea was that if you have an equivalence relation, there are ipso facto 'abstract objects' filling the bill for the phrase: *that which equivalent objects have in common.* Thus, for the equivalence relation *parallelism* in Euclidean Geometry, we have abstract objects: directions, for the equivalence relation *equidistance* between pairs of points in Euclidean Geometry, we have objects that are the distances between those points. Etcetera. One attractive feature of Frege's principle is that we can imagine the objects to be sui generis.[6] In contrast, e.g. equivalence classes in a set theoretical reconstruction are a *modeling* of the idea, but they carry the excess of a specific implementation.

The idea of abstracts being sui generis suggests that, always when we create abstracts, they form a type distinct from the already existing objects.[7] However, in the case that we consider

---

[4] Of course, this service is also performed, in different ways, by other approaches like Hintikka's Game Theoretic Semantics, Discourse Representation Theory and Martin-Löf's Type Theory.

[5] In Frege's Grundlagen ([Fre88]) the problematic relationship between Foundations of Mathematics and Philosophy becomes clearly visible. It is clear that we, as philosophers, want to understand what number is. On the other hand, it is hard to see why we cannot simply start with the natural numbers as a clearly understood structure in the foundational project considered as a project internal to mathematics.

[6] This is not Frege's idea but my preferred way of looking at it.

[7] The temporal locutions here are misleading: strictly speaking we want to have a conceptual order here.

numbers and extensions, we run into a problem, It seems undeniable that we can, in principle, count or collect anything, whatever it is. Thus, e.g., the number 3 can be also applied the concept $1, 2, 3$, and, thus, the idea of the type of the abstracts being brand new seems to founder here. Surprisingly, Frege's abstraction principle as applied to equinumerosity of concepts – this is called Hume's principle – is paradox-free, when taken on its own. (This can be shown by an easy argument that goes back to Peter Geach.) However, when we bring in extensions-as-objects, as is done in Frege's Axiom V, the Russell Paradox strikes.

I guess one should consider the state of affairs outlined above as a *skandalon* of Philosophy. Thus, a careful reflection on all steps of the argument is in order. The Neo-Fregean school and Kit Fine did just that. The technical side of the work of the Neo-Fregeans has been presented in lucid detail by John Burgess in a wonderful book [Bur05]. I think that, even if the conclusion of Burgess' book is that a Fregean program that does not import substantial further non-Fregean principles does not go very far[8], the path we take towards this conclusion is illuminating.

As so often with these things, the Neo-Fregean program has features that are very interesting from a more internal logical point of view. One of these is the precise determination of the strength of various predicative systems. Mihail Ganea showed in his paper [Gan06] that the first level of the predicative Frege hierarchy is mutually interpretable with Robinson's arithmetic Q. I showed that the $n+1$-th level of the hierarchy is equivalent to Q plus the n-fold iterated consistency of Q. See [Vis06].

What is the proper role of philosophy of mathematics in relation to logic, foundations of mathematics, the traditional core areas of mathematics, and science?

I have a bit of a problem with the question. Look. In a sense there is not really something like *philosophy of mathematics*. In the end there is just us trying to understand what mathematics is, asking these weird and probably incorrect questions, like *what are numbers?*. These strange questions could lead to rather strange answers. But if they do, they do. So, suppose e.g. someone says *the*

---

[8] I am conveniently assuming here that we need something of the strength of ZF for mathematics. Of course, there is a delicate discussion here. After all, you can already do a lot in Elementary Arithmetic.

philosophy of mathematics should respect mathematical practice.
I'd have to say that I do not really understand this idea. Say my reflections lead me to some version of finitism. Then, integrity dictates that I honestly and seriously think that idea through. I cannot imagine saying *this does not respect mathematical practice, so I'll drop the idea.* This just is not how the quest for insight works.

Conversely, I am happy that Mathematics is rather robust w.r.t. to our philosophical ruminations. It would not be a good idea when mathematicians changed their practice every time a new philosophical idea emerged.

So I'd like to say: there is no proper role. There are just a lot of questions —say about the nature of the continuum— and a lot of arguments and considerations. These questions, arguments and considerations may or may not interact fruitfully with logic, foundations of mathematics, the traditional core areas of mathematics, and science—and also computer science, linguistics and general philosophy. However, we cannot and should not *aim at* fruitful interaction. All we should do is honestly and seriously pursue the questions.

Well, maybe what I am saying above is too strong. Of course, there *are* examples of philosophical research explicitly designed to play a role in mathematics. I guess the great foundational programs are examples. However, I would like to distinguish Foundations from Philosophy proper—as was done in the statement of Question 3. Foundations of Mathematics is, at least in part, an internal mathematical endeavor. On the other hand, it has undeniably a substantial philosophical component. This component will also be a part of the Philosophy of Mathematics. So, perhaps, in this special case, we find a proper role.

What do you consider the most neglected topics and/or contributions in late 20th century philosophy of mathematics?

My choice would be: the Philosophy of the Variable and the Philosophy of Syntax.

I think Frege's *Entmythologisierung* of the Variable just was too successful. In the first place, there is the problem of reconstructing our intuitions of dependency of variables. In the second place there is the problem of thinking of variables in a 'coordinate free' and simultaneously compositional way. In the third place there is the question of kinds of occurrences of variables in a text and the connected problems of linking and coordination.

Kit Fine in his [Fin86] tried to redress the error of history. However, in the book he addressed just one of the problems: the problem of dependency. He did not study the side that I would call the dynamics of the variable: the philosophy of linking and of coordination. He did address that last problem in his inspiring John Locke lectures of 2002/2003.[9]

It seems to me that there is a whole vague cloud of ideas and considerations connected to the question of the nature of the variable which never converge to the clear picture one would like to see —if only we could see a bit clearer ... There are Frege's considerations on the Evening Star versus the Morning Star, Kripke's Pierre Puzzle, Kaplan on Words ([Kap90]), Marcus Kracht on Syntactical Structure[10], the use of discourse referents in Discourse Representation Theory ([KvE97]), registers and addresses in Computer Science, the underlining technique in Term Rewritting, syntax on sharing graphs in graph rewriting, the slowly emerging science of *the version problem*, treatment of the variable in Category Theory, occurrences of variables in Dynamic Predicate Logic, etcetera.

I guess one of the obstacles to progress is that our knowledge of Predicate Logic obscures connections. For example, in Predicate Logic we have constants and variables. The meaning of the variable in Predicate Logic is a function from assignments to objects.[11] It thus only depends on the domain of the model. The constant is more variable: for each model there is a specific choice of its value. The danger here is that we tend to think of names as analogous to constants and of anaphors as analogous to variables-as-treated-in-predicate-logic, where in reality both names and anaphors need a treatment different from both the treatment of constants and the treatment of variables we know from Predicate Logic.

The problem of Syntax is in a sense the same as the problem of

---

[9] See: ⟨http://www.philosophy.ox.ac.uk/misc/johnlocke/⟩.

[10] See the manuscript *The Emergence of Syntactical Structure*: ⟨http://www.linguistics.ucla.edu/people/Kracht/html/public-mathling.html⟩.

[11] If you say: ... *and an assignment is a function from variables to objects*, we seem to get into a circularity. There is a simple solution for this: we distinguish between the variable proper and its tag. It's the tags that are part of the assignments. But now we get into a typical Fregean problem as articulated in the beginning of [Fre75]: what are these conventional tags doing at the level of meaning? Surely logic is not about our arbitrary conventions?

the Variable. It seems to me that problems concerning coordination, coordinate freedom and compositionality are really general problems concerning syntax and not just concerning the variable. It's just that these problems come out so saliently if one tries to think about the variable.

What are the most important open problems in the philosophy of mathematics and what are the prospects for progress?

I hope that one more round of progress is possible for the problem *what is a number*. I really do believe that the natural numbers of daily life are cardinals, not ordinals. Of course, if we count we need to choose an ordering. But, similarly, if we specify a set of objects, one by one, we need to do so in an order. We understand, however, that this order is an artifact of the presentation, not an inherent feature of what is presented. So, also the order used in counting is an artifact of the process of determining number not a feature of number. I already mentioned the problems in defining cardinals along Frege's line. Still there is the lingering feeling that maybe there could be an idea that we are missing. I have no idea of what the chances of progress are.

The other problem, is the problem about the Variable and the Nature of Syntax. Here I think that we will see some real progress. It seems to me that many ingredients of the answer are just lying around. In a sense, we just need to get clearer about the question.

References

[BE87] J. Barwise and J. Etchemendy. *The Liar, an essay in truth and circularity*. Oxford University Press, New York, Oxford, 1987.

[Bur05] John Burgess. *Fixing Frege*. Princeton Monographs in Philosophy. Princeton University Press, Princeton, 2005.

[Fin86] Kit Fine. *Reasoning with arbitrary objects*, volume 3 of *Aristotelian Society Series*. Basil Blackwell, 1986.

[Fre75] G. Frege. Über Sinn und Bedeutung. pages 40–65. Vandenhoeck and Ruprecht, 1975. also reprinted in e.g. [Har94, pp. 142–160].

[Fre88] G. Frege. *Die Grundlagen der Arithmetik*. Felix Meiner Verlag, Hamburg, 1988.

[Gan06] M. Ganea. Burgess' PV is Robinson's Q. Unpublished manuscript, 2006.

[GS91] J. Groenendijk and M. Stokhof. Dynamic predicate logic. *Linguistics and Philosophy*, 14:39–100, 1991.

[Har94] R.M. Harnish. *Basic Topics in the Philosophy of Language*. Harvester Wheatsheaf, New York, 1994.

[Hol97] M.J. Hollenberg. An equational axiomatisation of dynamic negation and relational composition. *Journal of Language, Logic and Information*, 6(4):381–401, 1997.

[HV99] M. Hollenberg and A. Visser. Dynamic negation, the one and only. *Journal of Language, Logic and Information*, 8(2):137–141, 1999.

[Kap90] David Kaplan. Words. *Aristotelian Society*, Supp. 64:93–119, 1990.

[Kri75] Saul Kripke. Outline of a Theory of Truth. *Journal of Philosophy*, 72:690–712, 1975.

[KvE97] H. Kamp and J. van Eijck. Representing discourse in context. In J. van Benthem and A. ter Meulen, editors, *Handbook of Logic and Language*, pages 179–237. Elsevier, Amsterdam & MIT Press, Cambridge, 1997.

[MS76] Z. Manna and A. Shamir. The theoretical aspects of the optimal fixedpoint. *Siam Journal of Computing*, 5:414–426, 1976.

[Vis84] Albert Visser. Four-valued semantics and the liar. *Journal of Philosophical Logic*, 13:181–212, 1984.

[Vis89] A. Visser. Semantics and the liar paradox. In D. Gabbay and F. Guenthner, editors, *Handbook of Philosophical Logic, Topics in the Philosophy of Language*, volume IV, pages 617–706. Reidel, Dordrecht, 1989.

[Vis97] A. Visser. Dynamic Relation Logic is the logic of DPL-relations. *Journal of Language, Logic and Information*, 6(4):441–452, 1997.

[Vis98] A. Visser. Contexts in Dynamic Predicate Logic. *Journal of Language, Logic and Information*, 7(1):21–52, 1998.

[Vis01] A. Visser. On the ambiguation of Polish notation. Artificial Intelligence Preprint Series 26, Department of Philosophy, Utrecht University, Heidelberglaan 8, 3584 CS Utrecht, http://preprints.phil.uu.nl/aips/, July 2001.

[Vis02] A. Visser. The donkey and the monoid. Dynamic semantics with control elements. *Journal of Language, Logic and Information*, 11(1):107–131, 2002.

[Vis03] A. Visser. Context modification in action. Artificial Intelligence Preprint Series 43, Department of Philosophy, Utrecht University, Heidelberglaan 8, 3584 CS Utrecht,
http://www.phil.uu.nl/preprints/aips/, 2003.

[Vis04] A. Visser. Semantics and the liar paradox. In D. Gabbay and F. Guenthner, editors, *Handbook of Philosophical Logic*, second edition, volume 11, pages 149–240. Springer, Heidelberg, 2004.

[Vis06] A. Visser. The predicative frege hierarchy. Logic Group Preprint Series 246, Department of Philosophy, Utrecht University, Heidelberglaan 8, 3584 CS Utrecht,
http://www.phil.uu.nl/preprints/lgps/, 2006.

[VV96] A. Visser and C. Vermeulen. Dynamic bracketing and discourse representation. *Notre Dame Journal of Formal Logic*, 37:321–365, 1996.

25

Alan Weir

Professor of Philosophy
University of Glasgow, UK

Why were you initially drawn to the foundations of mathematics and/or the philosophy of mathematics?

Probably the influence of my logic teacher as an undergraduate, Neil Tennant. I had also been influenced by neo-Hegelianism, which featured on the curriculum in Edinburgh in the 1970s. Part of the attraction, I think, of philosophy of mathematics was that studying it formed part of a rebellion against what I saw (not wholly unjustifiably) as the wooliness of the neo-Hegelians. Of course logic, formal and philosophical, was attractive for similar reasons, and indeed, the work of W.V. Quine, whom I started to study seriously when on an exchange trip with Neil at Dartmouth College. I started mugging up by myself on set theory, having no background in mathematics beyond high school. At the end of that decade I was doing graduate work in Oxford, supervised among others by David Bostock, and I got interested in neo-Fregeanism while attending his lectures. This has remained an interest. Back in Scotland I was soon in contact with Crispin Wright and Peter Clark (the latter not, of course, a neo-Fregean) at St. Andrews and contact with the philosophers of mathematics based at St. Andrews, most recently in the Arche institute, has been a regular stimulus and springboard for my work in philosophy of mathematics ever since.

What examples from your work (or the work of others) illustrate the use of mathematics for philosophy?

Well work in philosophy of mathematics itself inevitably must pay some attention to actual mathematical practice, the more the better really except that, as always in philosophy, there is a pay-off between immersing oneself in the 'first-order' discipline and immersing oneself in the philosophical debate not only in the philosophy of that narrow area, but in philosophy more generally. A weak background in philosophy is liable to lead to poor philosophising about whatever the subject may be: mathematics, literature, sport. Conversely the more one knows of the particular area, the less one is likely to overgeneralise from particular cases. Where one lies on this spectrum probably depends a lot on the contingencies of one's career path.

In my own case, the mathematics I use to illustrate, indeed to test, my ideas in philosophy of mathematics would probably be seen as forming a very thin base by jobbing mathematicians; perhaps it is. Thus one must always, in deference to Kant, pay due homage to $7 + 5 = 12$. Many of the deep problems which have exercised philosophers who pay attention to mathematics, going all the way back to Plato– fundamental epistemological and ontological problems, problems of application to the empirical world– arise at this very sort of basic level (or, more traditionally, in basic problems in geometry).

In addition to elementary equations and basic results in geometry, arithmetic or analysis, such as the fundamental theorems of those last two domains, I venture very gingerly up the first steps of the arithmetic hierarchy to consider sentences like Goldbach's conjecture and, emboldened, sometimes press a little further on up, all this to consider the implications of undecidability for philosophy of mathematics. I have recourse, of course, to Gödel's two incompleteness theorems, allude to the more 'natural' undecidables due to Paris and Harrington and have some interest in the technical results of provability logic. I also have an interest in results which flow wherever the conditions of application of Gödel's results fail, e.g. in ω-logic and more generally in stronger infinitary logics. Like most philosophers of mathematics, I focus a lot on set theory, for example on the implications of independence results which flow from the work of Gödel and Cohen. Perhaps this focus on set theory is distorting though one can hold that the great generality and modelling power of set theory (allied with the simplicity of its basic ideas) makes it important even if it is not the only branch of mathematics which can play that role. One need not be a set-theoretic imperialist, in other words, to think of set theory as an especially important and fruitful area for philosophers to consider.

Probably the results which have most relevance for my own investigations into philosophy of mathematics are fairly simple, but incredibly important ones: Russell's paradox and the (arguably) related indefinability of truth result of Tarski (see below, point 5).

Turning to the use of mathematics in philosophy in general, the pre-eminent example of the use of mathematics in recent philosophy is surely the use of model theory, thus a form of applied set theory, and of metamathematics – proof theory – in studying, separately or together, formal languages and the notion of logical consequence for such languages. Of course the relationship between formal languages and the natural languages which we actually think in, when doing philosophy– or, indeed, when doing virtually all mathematics, including model theory and proof theory—is a difficult and contentious one. Even if, however, one has a highly negative view of the application of formal results to the systematic study of our grasp of natural language (and I, in fact, do not), one has to know the results in order to arrive at a well-grounded view on this.

Looking at things more broadly and historically, surely mathematics has been a central concern for so many great philosophers because of the problems it poses for empiricist and naturalistic philosophical viewpoints, as remarked above. Naturalistic philosophers have to take mathematics very seriously, since science seems so thoroughly immersed in it. (Hartry Field, of course, has tried to argue that this immersion is not strictly essential). But both the apparent ontology of mathematics, abstract entities outside of space and time (even 'less' physical, then, than Cartesian souls) and some of the methods of mathematics, in particular the apparent reliance on intuitively self-evident axioms, seem utterly at variance with what empiricists and naturalists tell us are legitimate.

Mathematics threatens to destroy naturalism from within, in other words. Forming a view on whether it does or not, is thus critically important for anyone interested in metaphysics (and indeed epistemology). Hence a well thought out of philosophy of mathematics is really essential for any serious philosopher. (From which, by the light of natural reason, it follows that philosophers of mathematics ought to be the best-paid of all philosophers, doesn't it? Alas, the light of natural reason shines but dimly in the corridors of academic power!)

What is the proper role of philosophy of mathematics in relation to logic, foundations of mathematics, the traditional core areas of mathematics, and science?

Many philosophers of mathematics urge us to take a deferential stance with respect to mathematicians and scientists. In particular, they argue that if philosophical considerations lead us (like Berkeley) to challenge the coherence or intelligibility of some mainstream part of mathematics (like classical analysis) or of, say, sub-atomic physics, then it is far more reasonable to suppose that there is a flaw in the philosophical reasoning than to conclude that we should cast analysis or sub-atomic physics into the flames as sophistry and illusion. We should go the tollens rather than the pollens route, in other words. In general I would agree with this sentiment, if non-dogmatically put. I do not myself advocate any serious revisionism of the sort envisaged by the intuitionists. But, as it were, I defend to the death (well maybe not that far!) their right to say what they say. That is, I do not think we should rule out of court the idea that there might be some deep flaw in orthodox mathematics and I do not think we (philosophers of mathematics that is) should view ourselves as mere underlabourers. We should not sell ourselves too short: there is deferential and there is doormat. Any half-way reflective mathematician or scientist will in fact have a philosophy of mathematics or of science, even if they do not realise it. These philosophical views often, to philosophers, seem to be ill-thought out or dependent on crude philosophical views long superseded. We have an intellectual duty not to adopt a doormat attitude but to challenge such views.

Furthermore, although I would not go as far as the Quineans in seeing no very determinate distinction between first-order mathematics (or science) and the second-order philosophy thereof, I think that mistaken philosophical views can have an impact on actual mathematical practice, which can thus be challenged. Thus if one rejects platonism then one may well eschew espousal of excluded middle (and, relatedly bivalence) for undecidable theses of the system one works with (e.g. the GCH for ZFC systems) even if one is not an intuitionist, and even if one accepts classical reasoning as at least pragmatically justifiable in 'everyday' areas of mathematics. (It is a tricky matter, of course, to try to implement an, as it were, restricted application of classical logic without spiralling off into radical revision.)

What do you consider the most neglected topics and/or contributions in late 20th century philosophy of mathematics?

I am not sure I can give a very objective answer to that question. From the perspective of my own particular interests, I would say pre-Hilbertian formalism and naïve set theory. Hilbert's formalism has deservedly received attention, for sure, but the earlier ideas are generally thought to have been smashed to smithereens by Frege. Similarly, Russell (and Zermelo) are conventionally thought of as having done for naïve set theory, though the work of Graham Priest on dialetheism (another strong influence on me, though one which pulls in opposing directions) has put a few much-needed dents into the bodywork of the orthodox position.

What are the most important open problems in the philosophy of mathematics and what are the prospects for progress?

Here again I will answer largely with regard to my own personal interests, without trying to take a more Archimedean view of the subject as a whole.

As remarked, one programme I have been interested in concerns the prospects for reviving pre-Hilbertian formalism, a formalism closer to the 'game formalism' of thinkers such as Thomae and Heine, so witheringly assaulted by Frege. I do accept that Frege's criticisms are for the most part justified. In particular, if formalism is the idea that mathematical utterances lack truth-evaluable content, form part of a mere syntactic game, then it is a hopeless position, unable to account for our theoretical knowledge of the syntax of these 'games' nor of their applicability in empirical science.

In recent decades, however, there has been interest outwith philosophy of mathematics in the idea that some forms of utterance can behave very much like truth-valued assertions but lack genuine ontological commitment, can fail to represent the world as containing certain objects, or features, or objects-with-features in the way that, for example, utterances in empirical science are paradigmatically supposed to do. I have in mind here, for example, projectivism about taste and some treatments of fiction. The most interesting and plausible forms of projectivism arise when one combines it with some form of deflationism to reach the conclusion that, since the utterances in the discourse in question walk, talk and quack like truth-valued assertions, they *are* truth-valued assertions. (This form of projectivism avoids the 'Frege/Geach' problem of explaining the validity and rational compulsion of inferences involving mixtures of projective and non-projective utterances.) The hope here is that one can have one's cake and eat it, one can proclaim truths about the aesthetic value of pieces of music or literature without ontological commitment to objective properties of beauty etc., and proclaim truths about Edmund Waverley or Dimitry Karamazov without commitment to the existence of the individuals described in the novels by Scott and Dostoyevsky.

Now while some have been tempted by the idea of a projective account of mathematics, I believe this to be a non-starter. The analogy between mathematical truth and 'fictional truths' such as 'Dimitry Karamazov had two full brothers' is, to my mind, stronger (which is not to say that one is led to fictionalism of the type espoused by Hartry Field). The key idea I have argued for is that utterances can be made true in some other way than by representing the disposition of objects and properties in some mind-independent domain, that there can be a 'non-representational' mode of assertion, with correctness in such assertion yielding non-representational truths. In mathematics, the most promising line of approach takes us back to game formalism: what makes mathematical utterances true is the existence of proofs, what makes them false, the existence of disproofs, though the content of those utterances does not, in general, involve reference to proofs and disproofs (any more than an aesthetic evaluation of Beethoven's late string quartets makes any reference to my internal attitudes and preferences).

Among the many problems facing such an approach probably the most telling ones flow from the following point. If we are to have any ontological gain over platonism then the proofs whose existence grounds the truth of mathematical utterances must be concrete proofs. But, given the limitations of human capacities, and finitude of human life, there will only be finitely many of these, with a finite upper bound on complexity of proof. Hence one seems driven to strict finitism, unless one can idealise the notion of proofhood without lapsing back into platonism in proof theory. Idealisation of the messily fragmented body of concrete mathematical practice, actual and potential, in any non-problematic sense, is essential anyway if any systematic theorising is to occur.

One common form of idealisation among, e.g. constructivist

philosophers of mathematics, takes what one might call the supernatural route. One attempts to explain our grasp of mathematics by appeal to the supposed capacities of supernatural beings with no finite limits on their powers (though constructivists do usually place transfinite limits: debar consideration of beings who can construct proofs with e.g. ε_0 number of steps; it is unclear how they can have a non-question-begging, independently motivated, basis for this restriction). I think such idealisation is illicit, that it is not comparable to legitimate idealisations in natural science but more like the introduction of supernatural agencies in pseudo-sciences such as 'creation science'. The question then arises: is there a different way to idealise concrete mathematical practice which validates this modified formalism? I believe there is; developing such an account of idealisation is thus, for me, a crucial project in philosophy of mathematics.

These concerns do not seem to be all that widely shared, though Neil Tennant has made one of the most detailed attempts to tackle the problem of idealisation from a constructivist perspective. The second area of my personal interests is, however, one much more widely shared, namely the whole problem of the paradoxes. I believe the problems posed by the paradoxes still remain completely open. A couple of decades ago I think that would have been thought of as a crazy view by virtually everyone, but now this has been whittled down to just 'most philosophers'! Most, that is, still think that, though philosophers are still wrangling with no end in sight about nearly everything else, here is something which has been conclusively settled.

It is true that more careful historical scholarship has cleared up a number of myths about naïve set theory in particular. For instance, Cantor's *mengen* are clearly not naïve sets. (But what about his *inhalte*, his 'domains'? They look much more like naïve sets.) Nonetheless the rather complacent assumption that the problems raised by the paradoxes of naïve set theory were quickly sorted out, for instance, by Zermelo, is now, justifiably, under some considerable pressure.

One thing which has made a difference has been the terrible problems philosophers of set theory get into when trying to tell us what on earth (or elsewhere) set theory is about. On the face of it, it deals with the objects in a particular domain–the set of all sets; truth and consequence should be defined so as to include this domain. But on standard set theories, no such universal set exists. Merely re-spelling 'set' as 'universe', 'indefinite totality' or 'domain' in its first occurrence in 'set of all sets' does not get around the problem. Similar problems attend any attempt to state what the theory of the ordinals, the backbone of the iterative set hierarchy, is about. The set of all ordinals does not exist, according to standard theory (the Burali-Forti paradox) and the ordinals themselves lack an order-type, they expand indefinitely into the distance in much the same fuzzy way that the natural numbers do according to radical constructivists. We cannot talk of them as a 'completed totality'. But actually it is very hard even to state this restriction on our ability to 'encompass' the ordinals (or all sets) without violating it. The temptation, indeed, to construct well-orderings which 'go beyond' Ω, the (non-existent, allegedly) order-type of all ordinals, is felt even by working set theorists– 'mouse theory'.

A related pressure on the conventional view has come from the development of inductive accounts of truth by Kripke, Martin and Woodruff which hold forth the prospect of restoring to some extent a 'naïve' theory of truth structurally paralleling in many ways naïve set theory. Part of the interest in such accounts has been dissatisfaction with the assumption that Tarski had cleared up the (structurally rather parallel) semantic paradoxes by dint of his hierarchy of metalanguages, an assumption as complacent as the analogous one concerning set theory.

And indeed a resort to hierarchy is really the key to virtually all the conventional responses to the paradoxes of set theory. This is clearly true of type theory. But the same also holds for Zermelo set theory, even allowing that the formalisation of an explicitly hierarchical iterative conception of set only emerged rather slowly after Zermelo's initial 'one step back from disaster' formalisation of set theory in 1904/8, perhaps not fully until Gödel's work on Cantor's Continuum Hypothesis. Even so, once model theory enabled us to give rigorous systematic interpretations of set theory, there emerged Tarski's hierarchy, or something analogous, namely a hierarchy of ever stronger theories augmented by stronger and stronger axioms (e.g. of infinity), each 'looking back' to interpret the earlier and in which the 'universe' of all sets of the previous theory is seen merely to be a proper sub-set of a wider universe. We get a 'hierarchy' (but what manner of beast is this?) of 'universes', not one in the hierarchy being genuinely universal. Hierarchies of one sort or another have seemed ineluctable, given Tarski's work on the truth.

So in one sense, the undefinability of truth result is for me very central. I think that the apparent unavoidability of that result (analogues hold for infinitary languages, as Dana Scott showed, so that it has much wider application than Gödel's results) has blinded philosophers to how disastrous it is epistemologically, at least for anyone hoping to give a systematic and non-trivial account of our grasp of mathematical language. Consider soundness results. Sometimes it is objected that any such result is 'merely' bootstrapping, as compared, one may suppose, with relative consistency results which show the consistency of a stronger theory from within a weaker one. Clearly the latter type of results are not available in all cases, they come to an end, at some point. But when they do, a bootstrapping result, a demonstration of the consistency or soundness of, for example, some set theory, *from within that very theory,* would be no 'mere' result, but a highly important stability test. It would not, of course, give Cartesian certainty that we are free from error, for an inconsistent system will be able to prove its own soundness; but it is widely seen now that it is wrong to expect certainty even in mathematics. But such a bootstrapping result would nonetheless show that the system passes a test of internal coherence comparable to that set by the programme of naturalised epistemology. A theory which 'says of itself' that it is sound is preferable, ceteris paribus, to one which does not.

What we actually have, however, where we work within the confines of the undefinability result, are proofs of the consistency of theory T from within a stronger theory T*. This is, epistemologically, an absolutely worthless result. It is like trying to authenticate the testimony of someone of unknown character by recourse to the testimonial of a proven perjurer with a vested interest in having the testimony accepted. The validator is even less trustworthy than that which is to be validated. We are left with no response, *even in our own terms* to the radical sceptic about the coherence of mathematics (or indeed logic in general), no response except blind dogmatism, a very unsatisfactory position to be in if one is attempting to defend the rationality of mathematics and logic against its irrationalist enemies.

Fortunately Tarski's resolution of the semantic paradoxes, once virtually unassailed, is now increasingly under challenge (initially via inductive theories, but these cannot be fully satisfactory since they are so clearly hierarchical themselves). Tarski's thoroughly hierarchical resolution has much in common with Russell's type theory, for instance as sketched in the latter's introduction to the *Tractatus* (an introduction less than enthusiastically received, to say the least, by Wittgenstein). This 'hierarchialism' leads to further problems, for both semantic and set-theoretic paradoxes, because of the danger of *superparadox*. Thus Graham Priest has persistently, and correctly in my view, urged that hierarchical solutions to the paradox involve introducing new semantic or set-theoretic notions (e.g. 'determinate truth' or 'indefinite totalities') in terms of which new, strengthened, versions of the paradoxes can be stated. An example is the notion of hierarchy itself. If, for example, we work with a hierarchy of meta-languages, what language is the notion of 'hierarchy' to be expressed in? If one of the languages in the hierarchy, then paradox will re-emerge in that language. If not, then what language is it expressed in? Whichever it is, this is the language that we should be interested in, if we are interested in a systematic and general theory for (an idealised version of) the language we speak. But now we have no resources with which to handle the original paradox, since the concept introduced to do so does not apply to the language in which it is expressed.

Considerations such as these have led Priest to *dialetheism*, to the view that some propositions (for example that the Russell set belongs to itself) are both true and untrue. In order to avoid triviality, Priest tries to combine naïve set theory with a paraconsistent logic such as LP or perhaps some non-monotonic version of LP. The most attractive logical approach here, in my view, utilises familiar classical operational rules, rules sufficient to yield classical logic given standard structural principles on proof-architecture, but restricts those structural rules in some way to yield the weaker, paraconsistent logic. But can one take such a dialetheist position seriously?

I have argued elsewhere both that it is possible and worthwhile to argue against such a view, as against those who would simply cite Anselm's injunction that the heretics of logic must be hissed away, and that a successful case can be prosecuted against dialetheism, a case which can be stated in terms acceptable to the dialetheist. I would, indeed, go further. The dialetheist acceptance of true contradictions is an even worse intellectual disaster than simply throwing one's hands up in despair (as many theorists now currently do) in the face of the paradoxes and abandoning all hopes of attaining a general view of the workings of language, specifically of mathematical language. Moreover the paraconsistent logics Priest has worked with are very weak. Even in the explosive context of naïve set theory, they do not yield any substantive mathematics.

Here, then, lie the germs of a research programme. Can we find restrictions on the structural rules of classical logic which transform it into a logic in which naïve set theory is not only non-trivial but in fact consistent and yet in which substantial amounts of mathematics can be carried out; ideally, in fact, in which a closed semantic theory for the language of naïve set theory itself can be given? I believe this can be done, at any rate that is one programme which I have been trying to carry through.

26

Philip Welch

Professor of Mathematical Logic
University of Bristol, UK

Why were you initially drawn to the foundations of mathematics and/or the philosophy of mathematics?

A simple fascination with the Barber and other paradoxes. However what was decisive was the Cantorian diagonal argument and the very notion of different kinds of infinity; the conundrum of the Russell paradox that not every description could be said to delineate a *set* of objects. These struck me as wonderfully mysterious and fascinating, and as questions were far more interesting than anything else I had heard in mathematics - and this just stayed with me. I can remember vividly where I first heard about these different kinds of infinity: it was amongst the rows of vegetable seedlings I was helping plant in the large derelict Victorian kitchen garden of my school that the new mathematics master, fresh out of university, had been allowed to take over. In a series of grey, damp, English, autumnal Wednesday afternoons whilst planting rows of potatoes, cabbages &c. he told me the 'story' of Cantor, and Russell's discovery of the 'error' in Frege. I recall being initially very resistant to the diagonal argument, thinking it absurd that one could, even as a hypothesis for a *reductio*, posit a lisiting of all decimal numbers! I particularly enjoyed the human drama of Russell writing to Frege to inform him of his 'mistake' by way of the Russell paradox. (I was only 11 at the time.) I cannot think of a purer example of a crucial intervention of a teacher in a child's mental life.

What examples from your work (or the work of others) illustrate the use of mathematics for philosophy?

The question seems to ask for instrumental uses of mathematics for philosophy rather than mathematics providing the raw material for philosophers to analyse (of the latter, of course there is an abundance). So I shall treat it as such. In as much as discussion concerning predicativity and that part of logic is allowed to be considered philosophy, then clearly the work of Feferman, and the mathematical proof-theoretical analysis of sub-systems of second order number theory have major implications in that debate. An even more general truism is the impact that 20'th century *mathematical*logic (I am thinking of all the tools, the analysis of first order logic as developed over that period) has had on *philosophical* debate. The very terms, or language in which philosophers think or place their arguments in so many fields of analytical philosophy owes so much to those advances (just think of the notion of "structure" for the simplest example, which has a very mathematical presentation). A further example is Tarski's inductive definition of the satisfaction relation, or inductive definition of "truth-in-a-structure" if you will. This could be regarded as a pure solution by mathematical induction to a philosophically inspired problem in logic. And look at the contribution to philosophical argument inspired by it? One can regard Gödel's Incompleteness theorems themselves as results in number theory about first order deductive systems. This was a crucial mathematical intervention in the formalist debate..

Although these are examples of the use or utility of mathematics in philosophy of mathematics, logic and the like, they are perhaps too general, or too obvious, to count as interesting "examples" here. So I shall just suggest two rather particular case-studies where mathematics intervenes.

(i) In "*Multiple Universes of Sets and Indeterminate Truth Values*" [8], Martin asks whether it is too unsharp a notion of set that could lead to a lack of truth value for certain sentences in set theory (such as the Continuum Hypothesis). To address this question (which I take to be part of the philosophy of mathematics, or of set theory, rather than part of set theory proper) he formulates both a strong and a weak notion of set formation by iteration, and uses a *mathematical*argument (very similar to those of Zermelo's (1930) categoricity arguments to show that sufficient isomorphisms are entailed (under a strong conception), to conclude that, even if the notion of set is not unequivocally sharp, that nevertheless questions such as CH—do have a determinate truth value. If there is no structure meeting our requirements under this strong conception, we may work under a weak conception, and pick out a suitable universe, which is formed by sufficient amalgamation of structures that, *e.g.*, a 'plenitudinous platonism' may allow.

The mathematical argument is one to construct such isomorphisms or amalgamations (admittedly in an object-based account of set objects). Certain assumptions are indeed made, *inter alia*, about wellfoundedness of stages in these constructions, to yield what might be termed a *philosophical* conclusion on the determinacy truth values, or at the very least to provide an argument against certain nominalist or non-object based accounts of set objects.

(ii) *Mathematics in theories of truth.* Kripke's seminal paper "*An Outline of a theory of Truth*",[6], certainly does not play up the mathematical parts of the theory. Nevertheless constructing a least fixed point over the structure of arithmetic for example, requires an induction along an initial segment of the ordinals, at least in a standard presentation. There are some who dislike a Tarskian-hierarchical language/truth-definition approach to truth. However speaking as a set theorist to whom hierarchical constructions are meat and drink, a hierarchy of languages with T-predicates seems very natural! Indeed what is Gödel's constructible universe L but a truth predicate iterated along the ordinals? Hierarchical solutions seem not unnatural, and in any case pretty much inevitable given Tarski's theorem on the Undefinability of Truth.

The least fixed point in Kripke's scheme, using say Strong Kleeneian truth tables, is a complete Π_1^1 set of gödel code numbers. It is a mathematical result that such sets are (computably (1-1) reducible to) sets of integers representing winning strategies for the first player in two person perfect information games with open payoff sets. This affords the prospect of using the theory of monotone inductive definitions and the theory of open games to give a very attractive game theoretic construction of, and a different epistemic slant on, this least fixed point. This presentation is then one which avoids an overt use of an ordinal hierarchy. (Martin has explicitly written out a nice game for this, at least in the Kleene's strong truth tables scheme case [7]. Similar game theoretic charcaterizations can be given for other similar theories of truth (see [11]). Either way a strictly mathematical theory underpins the philosophical construction.

Whether that actually *affects* the philosophical argument in the Kripkean schema, is perhaps moot.

However I think it does affect matters when comes to the *revision theories of truth* of Gupta and Belnap. This is an approach to analysing self-reference in the form of liar sentences and the like, which involves a full distribution of truth values to sentences containing a T-predicate. This results in conflicting assignments, and so the distribution of truth values is "revised". This must be done transfinitely often through all the ordinals. As this revision sequence rolls out, certain sentences receive a stable, fixed, truth value from some point on. Various flavours of these theories occur in the literature, but the full theory espoused by Gupta-Belnap ([2]) requires one to average out anomalous or undesirable judgements on sentences, by firstly considering all possible sequences, that is by starting with any initial distribution of truth values, but also secondly, by allowing the use of any "limit rule" as how to assign truth values at limit stages in this process. (Some versions ensure a "full variance" by *requiring* that all possible rules be used.) However a *mathematical* analysis ([12]) can be used to show that the stability sets they consider, meaning those sentences that become stable from some point on in all such sequences are of enormous complexity. For the structure of arithmetic the stability set is complete Π^1_2 in the language of mathematical analysis. (Under full variance this must be raised to Π^1_3.) I think the philosophical implications to be drawn from this is that for the structure of mere arithmetic to have such complicated "revision theoretic truth sets" is a warning flag. (Indeed Π^1_3-Comprehension Schemes are beyond the current capacity of ordinal analysis of Proof Theory.) Moreover questions concerning categorical "super-liar" sentences, run very quickly indeed into independence phenomena in ZFC set theory. Is an analysis of truth over arithmetic really to become undecidable in ZFC? Without the mathematical analysis however this phenomenon would have remained hidden from view. Again this illustrates a use of mathematics in the philosophical arena.

What is the proper role of philosophy of mathematics in relation to logic, foundations of mathematics, the traditional core areas of mathematics, and science?

(i) Analysis of Mathematical entities.

One proper role is still the traditional one of examining the entities that make up mathematical thinking, at the most fundamental level.

(ii) Analysis of Mathematical concepts.

I include here some of the conceptualization that occurs in the 'problems' at Qu. 5. below. I think philosophy of mathematics has a forward active role it should assume: in the same way that philosophical thought has entered into discussion with the very subject matter of quantum theory, evolutionary theory, &c. and closely examines the conceptual thinking there, so philosophy of mathematics should, hands-on, examine notions of proof, randomness, 'algorithm' etc (the kind of topics mentioned under Q5). The 'big-picture' concepts and debates remain, but philosophical analysis can, or could, concentrate on specific issues. This requires mathematical sophistication of course, but no more so than that of philosophers of physics, biology, and so forth have evinced.

(iii) Analysis of Mathematical activity.

I am not great a believer in that fuzzy object: 'the mathematical community'. Hence I find that the 'descriptive' character of Naturalism in the philosophy of mathematics disappointingly unengaged. I also do not think that the sum total of mathematical activity (as at 20.x.2007, 13.28 GMT) alone reflects what mathematics is. Or that mathematicians are the best arbiters or judges, or at least spokesmen or women, of what they do. They are often too busy doing it.

(iv) About science and the cross-over of mathematics into science.

Of course there is the whole question of how our perception of the world is interpreted or formed through mathematics and mathematical models, (the applicability question) but perhaps a more focussed question, *e.g.*, is to why *separability* features highly in mathematical models in science (pseudo-Riemannian manifolds, Hilbert space—indeed why Hilbert space at all?) Is separability so important? In order for philosophy of mathematics to be *useful* in such an area, it is asking for a particular micro-examination of some specific features of the mathematico-physical world view.

What do you consider the most neglected topics and/or contributions in late 20th century philosophy of mathematics?

1) Wittgenstein's remarks on Gödel's Incompleteness Theorem (and justifiably so).

2) The role of strong axioms of infinity, and in particular the philosophical status of determinacy axioms I would not say is neglected, but it seems in danger of it! For almost 20 years now we have had fundamental results about the universe of sets V that "large sets" affect the status of small sets (meaning those in $V_{\omega+1}$ or $V_{\omega+\omega}$). Results now show that such sets give radically new views about, *e.g.* definable subsets of the continuum. There has of course been debate, but it simply does not seem to have been given the attention it, not just deserves but, cries out for. I think this is not just a topic in philosophy of set theory but in philosophy of mathematics itself. If traditional mathematical analysis has been confined to the Borel regions of the universe, this of course has been because the methods in that tradition have been easily formulated as part of a, perhaps unconcsious, "ZC" *Weltanschauung* going back to Suslin, Lusin & co. We now see, and moreover *understand*, the connections between regularity properties such as Lebesgue measurability, the perfect subset property and the like, with representations of such sets as trees in large spaces with particular "homogeneity' properties, and in turn with elementary embedding properties of large submodels of the universe, or of the whole universe into a large submodel. One formulation, or mediation of this understanding is through *projective determinacy*, PD, which can be expressed as a first order statement in the hereditarily countable sets, and so about definable analysis. The status of PD has been written about by Woodin and others, but the fact remains that the metamathematizing of previous generations by adding large cardinal axioms, or stronger and stronger axioms of infinity to ZFC, on a somewhat *ad hoc* basis, is now part of a very *mathematical*picture of embeddings of submodels of the universe V. As Woodin writes: PD is compelling: every sufficiently strong axiom that we can conceive of at the moment when added to ZFC simply implies PD outright: this is *not* another syntactic forcing *consistency*proof. It means that in any sufficiently strong system $PD$is *unavoidable*. The way it arises is also part of an evolutionary picture: we may wish to assume that Gödel's L is *rigid* (that is there is no elementary embedding of L to itself which is not the identity map), and this gives a picture of the cardinality structure of V, and a lot of information can be gleaned about V *via* Jensen's Covering Lemma. However if we admit the possibility that L can embed to itself (which is not provable from ZFC, since this embedding property implies the existence of inaccessible cardinals in L and hence the consistency of ZFC itself) then

the situation is wide open: there is no natural line or apparent boundary to draw, no door in Bluebeard's Castle that should not be opened, before we get to the stage where we admit the possibility of embeddings that imply the strong axioms that prove *PD* outright. A naturalist account of the activity of set theorists seems inadequate to deal with this phenomenon of the unavoidability of *PD*, and indeed no philosophical account of the status of these conjectures about the embedding properties of V seems yet to come close. We cannot look back for help in discussions of how Zermelo or others took to the axiom of choice, or the axiom of replacement, or what then 'counted' as a good concept of set formation. We are in new territory. It is as if we have exhausted (at least for the moment) discussions on the 'iterative concept' of set, and now must move on to discussing the possibility that there could be elementary embedding properties of significant parts of the intuitive model, V, that the iterative concept delivers, and for which the ZFC axioms represent the formalised encapsulation.

What are the most important open problems in the philosophy of mathematics and what are the prospects for progress?

I should hesitate to assert any list of the *most* important problems, but should rather give *some* important, or at least I hope, interesting problems or areas, some speculative, some old.

(i) *Elucidation of the notion, or status, of proof in a machine age.* We have been faced with a small number of occurrences of long-standing mathematical theorems being *solved* by, essentially, *programming*. I am not considering here the notion of an unsurveyable (by single human) proof such as the classification of finite simple groups where the work was divided into pieces to be proven by different teams of people. I am thinking here of Appel and Harken, and the Four Colour Problem, where, as is well known, a large number of combinatorial cases were verified by, an albeit human programmed, computer. This has now been reworked in a (also in a computerized) formal proof program by Gonthier, that has been billed as the acceptable face of computer proof. Are traditional notions of proof as even (theoretically) surveyable to be augmented by the notion that we can correctly program a machine, be confident that the program performs correctly, and let it do the hard work? This is not a new question of course, but receives additional impetus from recent proof formalizers, such as those of

Gonthier above, and Avigad *et al.* Another more recent example is the work of Hales on the Kepler sphere packing conjecture [4], where a large number of inequalities again were machine-verified. This lead to great difficulties for the referees of the subsequent paper, even aside from the theoretical questions that arise. The notion of mathematics as a pencil and paper deductive activity is more severely challenged by some work on, Montgomery's conjecture and the Riemann zeta function, and L-functions generally, where computational numerical methods produce compelling evidence for various "theorems" concerning moments of L-functions at even greater distances from the origin. "Proof" is in danger of being abandoned in favour of powerful computation, albeit again allied with very great subtlety in mathematical argumentation.

(ii) *Elucidation of the notion of "algorithm"*. One would like to have to some acceptable, definitive, notion of general 'algorithm'. What is an algorithm? Is it a set of definable *recursors*? (as Moschovakis would have it; see [9]). Is it a *machine* or an *implementation* (Gurevich [3])? Moschovakis would argue that machine models are insufficient; and favours *recursors* (which he defines, and offers that they are related to recursive definitions much as differential operators are related to differential equations). However ("no entity without identity"!) he points out that recursor isomorphism is a stronger equivalence relation than algorithmic identity, and argues that we need a study of equivalence relations coarser than isomorphism to capture algorithmic identity.

(iii) *Elucidation of the notion of "randomness" in mathematics*. Naturally there is work on randomness as defined through Kolmogoroff complexity, and Martin-Löf randomness. The former has been much taken up by recursion theorists as a definition of randomness. However is it *the* definition of randomness? This seems to ignore notions of randomness that come from ergodic theory, that is, 'randomness' in dynamical systems, where 'truly random' is described as Bernoullian, and truly deterministic as being of 0-entropy (see Foreman [1]).

(iv) *Elucidation of the notion of "invariant"*. Mathematicians choose 'invariants' to characterise in some way various classes of objects. What is the status of 'invariants' and how do we know we have made a suitable choice of such? Of course in many cases the choice of an invariant is obviously canonical, there is some way of picking out a particular element of an equivalence class that serves as a representative or 'nominalizer' if you will, for that equivalence class. There are obvious echoes of Hume's Law from the Neo-Logicist school here: the process of abstraction from equivalence classes brings into being such nominalizers. However, moving away from the elementary arena of numbers as abstraction from equivalence classes of the same denumerable cardinality, abstraction principles for more complex mathematical relations look problematic. As evidence in this direction, for certain Borel equivalence relations it is impossibile to assign, in a Borel manner, invariants (I shall not go into an example, see the discussion again in [1]). One may adopt the ideology that "Borel", as a classification, is intimately tied to a countable amount of information, whilst "non-Borel" is inherently abstract. (As an example, the equivalence relation of isomorphism on countable linear orders is not Borel.) Or else turning the argument around, one says that one cannot live solely in the Borel world and expect to be able to find in any reasonable way not just 'typical' elements of any equivalence class from such relations, but one cannot expect to assign in any reasonable kind of way, a 'nominalizer' or 'name' or 'object' or 'invariant' to represent that equivalence class. What, if anything, does this imply for a neo-logicist program of constructing mathematics?

(v) Set theoretical axioms *versus* second-order logic. Why is it that second order logic or rather second order ZF^2, after having delivered Zermelo's wonderful categoricity results (see [14]) then delivers nothing else? Are there any relations at all between Woodin's *strong logics* arising from his work on the continuum problem, in the form of his Ω conjecture, and second order set theory? Woodin's approach (see [13]) seeks to factor out the baleful effects of Cohen's forcing method, by forming a *strong logic*, his Ω-*logic*, and seeks to find solutions, or at least approaches, to longstanding problems, such as the Continuum Problem (or parts thereof). The logic involves statements about all possible forcings at initial levels V_κ of the cumulative hierarchy, and relies on hypotheses of unboundedly many large cardinals to get off the ground. This logical consequence relation $\models_\Omega$ of the logic cannot be changed by Cohen's methods. The class of sets of reals needed to formulate the (non-finitary) provability relation $\Gamma \vdash_\Omega \varphi$ is that of the *universally Baire* sets of reals: this is a classification of sets of reals that extends in a very natural way, the Borel hierarchy. (The large cardinals are there to guarantee a sufficiency of such sets.) Woodin's Ω-*conjecture* amounts to asserting that completeness holds for this logic. This work involves hypotheses concerning the whole universe of sets at once, and although first-order in statement, one *might* speculate as to whether some form of second order formulation is somehow pertinent. Woodin points out that if the Ω-conjecture is true then Ω-logic is the logic of large cardinal axioms. (We cannot go into this here: see [5] for a discussion.) However if it fails, this could be because the set $\{\sigma | \models_\Omega \sigma\}$ is extremely complicated, it could be complete Π_2. However then it is recursively isomorphic to the set of integer codes of the logical validities in second order ZF (meaning in the full, non-Henkin semantics, see [10]). Apart from Koellner's very stimulating paper [5] there has been nothing (that I am aware of) written on this in the philosophical literature, (admittedly the technical machinery is daunting here.)

References

[1] M. Foreman. A descriptive view of ergodic theory. In *Descriptive Set Theory and dynamical systems*, volume 277 of *L.M.S Lecture Note Series*, pages 87–171. C.U.P., Cambridge, 2000.

[2] A. Gupta and N. Belnap. *The revision theory of truth*. M.I.T. Press, Cambridge, 1993.

[3] Y. Gurevich. Sequential abstract state machines capture sequential algorithms. *ACM Transactions on Computational Logic*, 1(1): 77–111, July 2000.

[4] T. Hales. A proof of the Kepler conjecture. *Annals of Mathematics*, 162(3): 1065–1185, 2005.

[5] P. Koellner. On the question of absolute undecidability. *Philosophia Mathematicae*, 2006.

[6] S. Kripke. Outline of a theory of truth. *Journal of Philosophy*, 72: 690–716, 1975.

[7] D. A. Martin. Revision and its rivals. *Philosophical Issues*, 8: 407–418,1997.

[8] D. A. Martin. Multiple universes and indeterminate truth values. *Topoi*, 20: 5–16, 2001.

[9] Y. N. Moschovakis. On founding the theory of algorithms. In H.G.Dales and G. Oliveri, editors, *Truth in Mathematics*, Oxford Science Publications, pages 71–102. O.U.P., Oxford, 1998.

[10] J. Väänänen. Second order logic and foundations of mathematics. *Bulletin of Symbolic Logic*, 7(4): 504–520, 2001.

[11] P. D. Welch. Games for supervaluation and dependency. *http://www2.maths.bris.ac.uk/emapdw/games3.pdf*.

[12] P. D. Welch. On Gupta-Belnap revision theories of truth, Kripkean fixed points and the next stable set. *Bulletin of Symbolic Logic*, 7(3): 345–360, Sep. 2001.

[13] W. H. Woodin. Set theory after Russell: The journay back to Eden. In *One hundred years of Russell's Paradox*, volume 6 of *Logic and its Applications*, pages 29–47. de Gruyter, 2004.

[14] E. Zermelo. Über Grenzahlen und Mengenbereiche: Neue Untersuchungen über die Grundlagen der Mengenlehre. *Fundamenta Mathematicae*, 16: 29–47, 1930.

27
Crispin Wright

Professor of Logic and Metaphysics
Wardlaw Professor
University of St. Andrews, UK

Why were you initially drawn to the foundations of mathematics and/or the philosophy of mathematics?

My undergraduate degree at Cambridge (and especially part IIB of the then-called Moral Sciences tripos) involved a concentration on formal and philosophical logic, and the philosophies of science and mathematics. It is unsurprising that such a curriculum, taught in a framework that included the inspiring supervision of Casimir Lewy has conditioned my philosophical interests ever since. But there are a number of other factors that have contributed to sustain this focus. It is no accident that very many of the fathers of contemporary analytical philosophy – Descartes, Leibniz, Kant, Berkeley, Frege, Russell and Wittgenstein – as well as great contemporary figures like Quine, Putnam and Dummett – have been intensely interested in the philosophy of mathematics. While the reasons for this are no doubt complex, two considerations seem to me to be of paramount importance.

The first is that, from a philosophical point of view, the very existence of *pure* mathematics, and its immensely fecund, even indispensable role in application to ordinary empirical thought and in science, makes it an especially challenging phenomenon. Mathematics was viewed as the paradigm of knowledge long before mankind had evolved any robust conception of empirical scientific method – to the Greeks indeed, the idea seems to have come quite naturally that the reach of pure mathematical thought might extend to include the most fundamental explanation of the nature of the world. If the modern mind has reserved that project for theoretical physics, it still remains that, according to our ordinary thinking, much can indeed be known about the world by pure mathematical means; indeed, physics itself would hardly be possible without the use of tools developed by the mathematician whose status as instruments of practical discovery is thus implicitly unchallenged. So it is thus a standing challenge to explain how mathematical techniques are at the service of knowledge, both within mathematics and beyond. The mystery deepens when one reflects that mathematical knowledge seems to be knowledge of things that are so non-contingently. And it deepens still further when one factors in its apparent special subject matter – an abstract realm of numbers, sets, points, planes, and functions. How is it possible to know of such things by pure thought? And how can the knowledge thereby gleaned be of relevance to the ordinary physical world?

I think it was the mix of the ready intelligibility of these questions with their difficulty and their evident importance – the sense one has that answers to them will profoundly condition one's conception of much more ramified issues in metaphysics and epistemology – that explains their initial fascination. A crucial additional stimulus in my own case was attending a course of lectures, in my second year as an undergraduate, on Cantorian set theory and experiencing, like so many others, a sense of intellectual awe at the hierarchies of the transfinite that Cantor seemed to have discovered. It was two years later, on reading Wittgenstein's *Remarks on the Foundations of Mathematics* for the first time, that I became aware of both of the possibility of an utterly different take on the matter, and also of how deep reaching are the hostages, in the areas of mind and meaning, of my original awe-struck reactions. It was that realisation, of the philosophical depth of issues in the philosophy of mathematics, which has done most to sustain my interest in the topic.

The second point is connected but consists in a kind of converse. Encountering Michael Dummett's writings towards the end of my time as a PhD student, I became aware of the extent to which fundamental issues and debates in other areas of philosophy often allowed of an especially focused formulation within the philosophy of mathematics, so that it could to some considerable extent be used as a testing bed for general ideas of importance in areas of philosophy quite remote from it. Dummett's own interpretation of the philosophy of the mathematical Intuitionists is an especially forceful example of this, offering as it does one construction of the opposition between broadly realist and anti-realist controversies about all kinds of discourse. A second very important example is provided by the debates about mathematical existence, in which the various protagonisms – platonism, nominalism, reductionism, fictionalism, and so on – have counterparts that engage, once again, in discourse of almost every kind. There is of course no presumption that these various debates permit a uniform resolution, or that the strengths of the various positions in one area of discourse tie in closely with those they have in another. Exploring the differences is part of what makes the philosophy of mathematics such a fruitful starting point.

What examples from your work (or the work of others) illustrate the use of mathematics for philosophy?

Had the question concerned the "use of mathematics *in* philosophy" I would be sceptical that any significant examples could be cited. There are philosophers – recently, and somewhat notoriously, Timothy Williamson, for example[1] – who would welcome an increased use of, for example, model-theoretic techniques in the way that philosophical issues are discussed. But of course the interest of the results yielded by such methods will depend on the extent to which the models explored adequately capture the concepts under philosophical investigation – something whose determination must necessarily fall back on less formal techniques of reflective characterisation. A good example of the point is provided by Church's thesis, that the effectively calculable arithmetical functions are exactly the general recursive ones. General recursiveness is certainly a mathematical notion, and – on the assumption that the thesis is true – we can learn a lot about the nature of effective calculability by mathematical exploration of the notion of general recursiveness. But mathematics cannot teach us that Church's thesis is true. The proposal that it is is an essentially philosophical conjecture, comparing an informal intuitive notion with a technical one, and as such is beyond verification by formal techniques. It is characteristic of philosophical analysis, even when, as in this case, the analysans draws on technical resources, to be informal and conjectural in this way.

When the question concerns the use of mathematics *for* philosophy, a much more expansive assessment is possible. Gödel's incompleteness theorem stimulated a reinvigoration of the debates

---

[1] See his "Must Do Better" in P. Greenough and M. Lynch, Eds., *Truth and Realism*, Oxford University Press, 2006.

about mechanism in the philosophy of mind. It provided both for sharp formulation of the terms of the debate, and for one famous (widely disbelieved) argument[2] that opponents of mechanism continue to run to this day. John Wiles' resolution, at last, of the status of Fermat's "Last Theorem" is beginning to provoke philosophical discussion of the status of proofs that deploy, seemingly unavoidably, more advanced branches of mathematics in the solution to problems posed by simpler ones. Skolem's theorems put the awe-struck frame of mind, epitomised in the phrase "Cantor's Paradise", under severe philosophical pressure.[3] And, in my own work, the re-discovery of what has come to be known as "Frege's theorem", that the Dedekind-Peano axioms allow of derivation in a system of second order logic with Hume's Principle as sole additional axiom, has enabled us – as might otherwise have been impossible – to focus on at least some of the profound issues in the basic epistemology of logic and elementary arithmetic that would have been very salient much sooner had the contradiction in Frege's system of *Grundgesetze* never been discovered.

Uses of mathematics for regions of philosophy other than the philosophy of mathematics are less salient. But one very nice example is provided by an observation of Stephen Schiffer. James Pryor's influential epistemological 'Dogmatism' offers an account of the confirmation of perceptual judgements by observation of which it is a central feature that, for suitable contents, "P", an experience as of its being the case that P raises the probability of P independently of any presupposition about the subject's collateral information. This proposal promises to sustain a version of G.E. Moore's so called proof of an external world. If my experience as of my having a hand in front of my face unconditionally raises the probability of my actually having a hand, than it does the same for the consequences that there is an external world (since a hand is a material object, existing in space), and that I am not a disembodied brain in a vat. But according to classical probability theory, when two hypotheses, H_1 and H_2, are each consistent with evidence E, then that evidence can raise the probability of one at the expense of the other only in a context in which the respective

---

[2] Locus classicus: J. R. Lucas, "Mind, Machines and Godel", *Philosophy* 36 (1961) pp. 112-127. Lucas receives impressive support in Roger Penrose's *The Emperor's New Mind*, Oxford University Press 1999.

[3] See the symposium, "Skolem and the Sceptic", by Paul Benacerraf and Crispin Wright, *Proceedings of the Aristotelian Society*, supplementary volume LIX (1985)

prior probabilities, of H_1 on E, and of H_2 on E, reflect the disparity. If the prior probabilities in turn have to be based on evidence, then Priors proposal in epistemology is inconsistent with classical probability theory.[4]

What is the proper role of philosophy of mathematics in relation to logic, foundations of mathematics, the traditional core areas of mathematics, and science?

There is, of course, no reason why the philosophy of mathematics should have a uniform role in relation to these various disciplines and areas of concern. As far as the traditional core areas of mathematics – arithmetic, analysis, set theory and geometry – are concerned, perhaps the simplest way of captioning the principal role of the philosophy of mathematics is to see it as centred on the instance, local to these areas, of what Christopher Peacocke has usefully termed the "integration challenge".[5] The integration challenge raised by any theory, or area of discourse, is to provide an account of the nature of the subject matter concerned which reconciles it with a satisfactory local epistemology – a satisfactory account of the possibilities for knowledge in that area, dovetailing smoothly with what we take to be our actual methods of knowledge acquisition about it.

Integration presents a challenge because, especially in the philosophy of mathematics, the straightforward approach to the ontological issues characteristically gives a rise to epistemological dissatisfaction, and conversely. Paul Benacerraf's epochal "Mathematical Truth" gives vivid expression to this.[6] Platonism about numbers and sets offers an account of the subject matter of mathematics which has attractions when one is preoccupied by its apparently necessary, non-empirical character; but the price one pays, as philosophers from J.S. Mill to Hartry Field have emphasised, is to obscurantise the role of mathematics in scientific theory, and to call in question our competence to know the propositions of pure mathematics that we standardly take ourselves to know. Field's own notorious response[7] is to deny that mathematics *has* a proper

---

[4] James Pryor "The Skeptic and the Dogmatist", *Noûs*, Volume 34, Number 4, December 2000, pp. 517-549; Stephen Schiffer, "The Vagaries of Skepticism" *Philosophical Studies* 119 (2004), pp 161–184

[5] Christopher Peacocke, *Being Known*, Oxford, Clarendon Press 1999

[6] *The Journal of Philosophy*, 70, No. 19, pp. 661-679

[7] H. Field, *Science Without Numbers*, Princeton University Press 1980

subject matter, and to recast the issues about its epistemology precisely in the setting of its utility in science. Constructivist, and formalist proposals, though very different in detail, may be viewed as likewise prioritising the demands of a workable epistemology of mathematics. If the subject matter of mathematics could somehow be made out to be the very symbols we manipulate in the course of doing it, or the very proofs we construct, then it might seem as though there should be no problem about explaining how the methods of mathematics engage with its subject matter. The problem is acute. I believe that the most promising extant approach to providing an account of the central theories of classical mathematics which explains the nature of their subject matter, our ability to know the truths of their standard axioms, and how the latter may be a priori yet carry a content fitting them for their standard applications, is the *abstractionist* programme that has been championed by Bob Hale and myself. But I will not attempt to further the defence of that claim here.[8]

In my view, the primary role for the philosophy of mathematics in relation to the foundations of mathematics is highlighted by Wittgenstein's impatient question, "What does mathematics need a foundation *for*?"[9] It was very much in the spirit of the logical atomism of Russell and Wittgenstein to suppose that something deeper had to underlie the, as it were, temporary resting points of classical mathematical theories in their normal axiomatisations. Frege's attempt to uncover the deeper basis having foundered in paradox, it then understandably came to seem that new foundations were wanted urgently – as if the suspension cables of a bridge had been found to be badly corroded, and collapse imminent unless they could be renewed. But we are no longer atomists, and no longer, most of us, subscribe to the idea that mathematical knowledge rests on, or requires, some deep basis which needs to be uncovered. So Wittgenstein's question can seem well taken. A primary concern for philosophy of mathematics should be to determine whether it is.

It will certainly seem so if one subscribes, as many do, to the broadly Quinean picture of the role of mathematics in empirical knowledge supported by writers such as Shapiro.[10] In that case,

---

[8] See Bob Hale and Crispin Wright, *The Reason's Proper Study*, Oxford University Press 2001.

[9] *Remarks on the Foundations of Mathematics*, V, 13

[10] Stewart Shapiro, *Philosophy of Mathematics: Structure and Ontology*, Oxford University Press, Oxford, 1997

mathematics no more needs a foundation than physics does. It participates, albeit in a specially fundamental role, in the totality of our scientific knowledge, is known in fundamentally the same ways as the propositions of science, and the axioms of its theories stand or fall under empirical pressure in the end, as witness the case of Euclidean geometry. What stands against this, of course, is the traditional picture of mathematics, or at least some mathematics, as both a priori and relatively certain. For those, and there are again many, who incline to this view, the need to provide an account of how mathematics enjoys this status is paramount. Such an account need not necessarily be technical, as it became in the hands of the logicists. But if mathematics is somehow epistemologically exceptional, there is no dodging the philosophical pressure to explain why.

The interplay between debates in the philosophy of mathematics and issues concerning logic has been one of the most interesting features of twentieth century philosophy of mathematics. Frege famously conceived of logic as a codification of the "laws of thought", a body of absolute, normative principles constraining all rational minds. He conceived of this view as standing opposed to psychologism, a loose name for a number of tendencies united in the idea that logic is somehow a merely descriptive science, concerned with the systematisation of human inferential propensities. But the gradual erosion of Frege's perspective through the twentieth century stemmed less from any psychologistic tendency than from a gradually emerging pluralism, a range of standpoints that retain the normativity of logic, for the most part, but have denied its absoluteness. Some of the impetus to this direction has originated, once again, in Quinean views about the alleged revocability of logic under the pressures of empirical theorising. But it is the Dummett-inspired debate between intuitionists and classicists in philosophy of mathematics that provides the most striking example.

For the Dummettian intuitionist, logic is still normative. And there is still such a thing as the *right* logic. But in stark contrast to anything Frege thought, *which* logic is the right one is allowed to vary with the demands of the subject matter to which it is applied. Classical logic is fine for reasoning involving only finite totalities and decidable properties. But revisions are demanded elsewhere. In particular, it is the intuitionist view that the proper philosophical account of the content of statements concerning the infinite demands a non-classical logic in which the laws of double negation elimination and excluded middle fail, and substantial qualifications are demanded to the interdefinability of the quantifiers, and the De Morgan laws. It is an open question to what extent arguments parallel to those of the intuitionists in mathematics may be developed for statements concerning the potentially undecidable future and past, and also for vague discourse. But it was controversy in the philosophy of mathematics that fomented the idea that logic might be revisable in a principled way on grounds other than empirical expediency while still retaining the Fregean conception of logic as essentially normative over correct thought. It is ironic that the classical reaction against this revisionism is largely motivated by the Quinean standpoint, which is quite antithetical to Frege's.

Comment on the interaction between debates in the philosophy of mathematics and issues concerning science, or the philosophy of science, had best be left to those with more than an amateur acquaintance with the latter. But let me close this section by flagging one issue which I would dearly wish to see better researched: the status of the often asserted claim that only classical analysis is fitted to serve the mathematical demands of contemporary physics. For the little that I understand about the issues, no clear reason is apparent to me why this should be so. But it will demand a very able philosopher with a rare grasp both of classical analysis and of various non-standard, including especially intuitionist accounts, together with an extensive grasp of theoretical physics, to set the record straight. Until someone does the orthodoxy that classical mathematics retains essential advantages in point of "simplicity, power, past success and integration with theories in other domains",[11] is likely to remain unchallenged simply because no one knows any better.

---

[11] These are actually Timothy Williamson's words about classical *logic* (Williamson, Vagueness Routledge 1994, p. 186) but they are perfect for the sentiment about classical mathematics too.

What do you consider the most neglected topics and/or contributions in late 20th century philosophy of mathematics?

The topic perhaps most obviously neglected, since it was so centrally on the agenda during the nineteenth century and before, is continuity. I have in mind specifically the cluster of issues to do with the application of real and functional analysis to the empirical world. The notions of, for example, a continuous change in height, or a continuous rise in temperature, are notions with, it seems, a relatively clear if rough and intuitive empirical content. What is the proper philosophical analysis of this notion, or – should there prove to be a family – notions, and how does it mesh with the various technical notions of continuity developed in classical analysis, non-standard analysis, and intuitionist analysis? Many of the issues here would surely have loomed large in the unpublished – and probably never written – parts of Frege's *Grundgesetze*, informed as it would have been by his constraint that a proper philosophical account of the concepts of a mathematical theory should somehow write in the potential for its canonical applications. Such an account would demand an analysis of the notion of quantity, and an exploration of whether anything could be said about the kinds of variation in magnitude that quantities of different kinds – height, masses, temperatures, directions, and so on – might be capable of displaying independently of the superimposition upon them of a particular mathematical conception. The question highlighted at the end of the preceding section would also belong to this agenda.

It's not completely clear why, this range of issues, crucial to understanding the role of mathematics in empirical science, should have proved so much less interesting to philosophers of mathematics than issues raised by set theory, the limitative theorems of Gödel and Skolem, and broad concerns about mathematical ontology. Frege's failure, before his programme crashed around his ears, to carry it far enough to demand engagement with them may have been a factor; an obsession with the paradoxes, and with the general epistemological and reconstructive issues which they raise, has probably been another.

27. Crispin Wright

What are the most important open problems in the philosophy of mathematics and what are the prospects for progress?

I confine attention to those areas of philosophy of mathematics where I have my own greatest investment, and where I think there are realistic chances of progress. First, then, on the philosophy of set theory: we have known since the paradoxes first broke that, if we are to continue to think of sets as in any significant way the objectification of properties, we are going to have to allow that some properties do not determine sets. The insight then required is to determine what it is about those properties that fail to determine sets explains their failure so to do. There are various traditional answers. But the one that comes closest to being intuitively satisfying, to my mind, is Russell's idea that the rogue properties are those which, in the later terminology of Dummett, are *indefinitely extensible*: properties such that purported quantification over all their instances subserves the definition of new instances which, on pain of contradiction, must lie outside the range of the original quantifiers. I believe that we are getting closer to a satisfyingly rigorous characterisation of this notion, and a consequent deepened understanding of why the paradoxes arise.[12]

Finally to the abstractionist programme. For all the inroads made, the relative neglect of two large issues has somewhat compromised the achievements of the programme to date. The first is the challenge of developing an account of the systems of second order logic in which abstractionist theories are characteristically framed to provide a suitable and satisfying alternative to the Quinean idea that they are in effect systems of set theory – a notion lethal to the interest of abstractionist proposals, which if the Quinean view were correct, would collapse into a rather baroque form of set-theoretic foundation. The other challenge is to find some principled distinction between those abstraction principles which, as one assumes is the situation of Hume's principle, are suitable to play a conceptually and epistemologically illuminating role in the foundation of mathematical theories and those – the principles in "bad company" – are not. The latter question has been worked on extensively recently and there are now clear signs

---

[12] This optimism is based on my experience of co-authoring "All Things Indefinitely Extensible" with Stewart Shapiro, published in Agustín Rayo and Gabriel Uzquiano (eds.), *Absolute Generality*, Oxford University Press, 2006

that, in one of various ways, a stable, positive account may not be far away.[13] Progress on the former will depend on a radical rethink of the nature of quantification and a deeper understanding of the sense in which it is an operation of logic. I believe that ideas tentatively introduced in my own recent work may provide significant steps towards the prerequisite notions.[14]

---

[13] See Oystein Linnebo, ed., special number of *Synthese* on the Bad Company Problem, forthcoming.
[14] C. Wright, "On Quantifying into Predicate Position: Steps towards a New(tralist Perspective" in Michael Potter, ed., *Mathematical Knowledge*, Oxford University Press 2007

28
Edward N. Zalta
Senior Research Scholar
Stanford University, USA

Reflections on Mathematics

Though the philosophy of mathematics encompasses many kinds of questions, my response to the five questions primarily focuses on the prospects of developing a unified approach to the metaphysical and epistemological issues concerning mathematics. My answers will be framed from within a single conceptual framework. By 'conceptual framework', I mean an explicit and formal listing of primitive notions and first principles, set within a well-understood background logic. In what follows, I shall assume the primitive notions and first principles of the (formalized and) axiomatized theory of abstract objects, which I shall sometimes refer to as 'object theory'.[1] These notions and principles are mathematics-free, consisting only of metaphysical and logical primitives. The first principles assert the existence, and comprehend a domain, of abstract objects, and in this domain we can identify (either by definition or by other means) logical objects, natural mathematical objects, and theoretical mathematical objects. These formal principles and identifications will help us to articulate answers not only to the five questions explicitly before us, but also to some of the other fundamental questions in the philosophy of mathematics raised below.

---

[1] This theory was outlined in detail in Zalta 1983 and 1988, and has been applied to issues in the philosophy of mathematics in the works referenced below.

Why were you initially drawn to the foundations of mathematics and/or the philosophy of mathematics?

As a metaphysician, I've always been interested in data that consists of (apparently) true sentences and valid inferences that appear to be about objects and relations other than those studied by the natural sciences. These sentences and inferences often form part of a correct description of the world, and the challenge for the metaphysician is to explain this data, by developing a systematic theory of the truth conditions for the sentences in question that reveal why those sentences have the apparent truth value, and consequences, that they do have. The sentences and inferences deployed in the practice of mathematics are interesting examples of this kind of data, since they appear to reference, or quantify over, special objects and relations that are not studied *per se* by the natural sciences. The data are made even more interesting by the fact that serious scientific investigation employs the language of some segment of mathematics. Thus, as a metaphysician, it is important to develop an overall ontological theory that allows us to assign a significance, or denotation, to the terms and predicates of mathematical sentences in such a way that accounts for the truth of those sentences and for the valid inferences that we may make in terms of them. This is what drew me to the philosophy of mathematics.

As to the foundations of mathematics, I shall adhere to the distinctions, drawn explicitly in Shapiro 2004, between the metaphysical, epistemological, and mathematical foundations for mathematics. Metaphysical foundations for mathematics address the issues outlined in the previous paragraph. Epistemological foundations for mathematics center around the questions: (1) what kind of knowledge is knowledge of mathematics?, and (2) how (by what cognitive mechanisms) do we acquire such knowledge? Mathematical foundations for mathematics address the questions: (1) Is there a mathematical theory distinguished by the fact that all other mathematical theories can be reduced to, or translated into, it? (2) What notions of reducibility and translatability are appropriate for comparing the strength of mathematical theories?

Now given these distinctions, I shall focus primarily in what follows on metaphysical and epistemological foundations for mathematics. As to mathematical foundations for mathematics, I shall assume that although philosophers may have something to say about the nature of the reducibility or translatability relation in play when determining the strength of mathematical theories, it is primarily a mathematical, and not a philosophical, question as to whether there is a foundational mathematical theory. Therefore I shall suppose that the metaphysical and epistemological foundations of mathematics developed by philosophers should not imply whether there is or is not a single foundational mathematical theory. The metaphysics and epistemology of mathematics should be consistent with whatever conclusion mathematicians (including set theorists, category theorists, etc.) draw with respect to the existence of such a theory. Some philosophers might wonder how this is possible, but the theory described below shows that it is. Finally, I should mention that there are many other non-foundational issues in the philosophy of mathematics, but I shall have little to say about them here.

What examples from your work (or the work of others) illustrate the use of mathematics for philosophy?

To answer this question, let me distinguish philosophical questions about the foundations of mathematics (i.e., the metaphysical and epistemological foundations discussed above) from other philosophical questions. In my view, philosophical theories about the foundations of mathematics should employ mathematical methods (e.g., the axiomatic method) but not assume any mathematically primitive expressions other than numerical indices to indicate the arity of relations.[2] I see metaphysics as an *a priori* science that is prior to mathematics: whereas mathematical theories are about particular abstract objects (e.g., the natural numbers, the ZF sets, etc.) and particular relations and operations (e.g., successor, membership, group addition, etc.), metaphysics is about abstract objects in general and relations in general. So metaphysics should be free of mathematical primitives, though primitive mathematical terms and predicates might be imported into metaphysics when those primitives are accompanied by principles that identify the denotations of the terms and predicates as entities already found

---

[2] This appeal to numerals to indicate the arity of relations doesn't entail, as Frege realized, that we quantify over numbers. One might eliminate the numerals by using tics, i.e., indicating the arity of relation F as F', F'', F''', ... But if it could be shown that there is a ineliminable appeal to the natural numbers in using numerals in this way, then it may be that the best we can do is use this numeralized logic of relations to reconstruct the concept of *number* and the Dedekind-Peano postulates from our metaphysical first principles, as in Zalta 1999.

in the background metaphysics. Another reason not to have mathematical primitives in our metaphysical foundations is to avoid ontological danglers. That is why set theory or model theory cannot serve as a metaphysical foundations; the metaphysical and epistemological problems about mathematics cannot be solved by an appeal to set theory or model theory, for that is just more mathematics and therefore part of the data to be explained. Such problems must be solved by an appeal to a more general theory of abstract objects and relations.

Thus, while I endorse and use the axiomatic method to organize a metaphysical foundations for mathematics, that framework (the 'second-order' modal theory of abstract objects),[3] employs only the following metaphysical and logical primitives: *individual* $(x, y, z \ldots)$, *n-place relation* $(F^n, G^n, H^n, \ldots)$, *exemplification* $(F^n x_1 \ldots x_n)$, *encoding* (xF^1),[4] *it is not the case that* $(\neg \phi)$, *if-then* $(\phi \to \psi)$, *every* $(\forall \alpha \phi)$,[5] and *it is necessarily the case that* $(\Box \phi)$.[6] In the higher-order formulations of the theory of abstract objects, we sometimes also employ the notion of *type*, defined recursively in the usual way. As we utilize this framework for the philosophy of mathematics, the fact that it is mathematics-free should be kept in mind.

So mathematical methods, such as the axiomatic method, may be used in responding to philosophical questions about the foundations of mathematics. But, of course, there are many other philosophical questions that have nothing to do with the foundations of mathematics. Here, the philosopher is free to employ whatever mathematics suits the task at hand. Some early examples are very well known. Leibniz (1690) used an algebraic operation ($\oplus$, for concept addition) and axioms for semi-lattices (governing the relation $\preceq$ of concept inclusion) to formulate his 'calculus of concepts'. Frege (1891) employed functions and functional application (conceived mathematically) to analyze predication in natural language. The 20th century saw an explosion of such applications

---

[3] I put 'second-order' in quotes because while the language of the theory is second-order, the theory doesn't require full second order logic.

[4] See Linsky & Zalta 2006 (80) for a discussion as to why the new form of predication, x encodes F (xF) is not to be conceived as a mathematical primitive.

[5] Here α may be any individual variable or relation variable.

[6] The theory also employs a distinguished predicate '$E!$'. Formulas of the form '$E!x$' and '$xE!$' are to be read as 'x exemplifies being concrete' and 'x encodes being concrete', respectively.

of mathematics in philosophy. A noteworthy recent example is Leitgeb's use of the (graph-theoretic) mathematics of similarity relations to understand Carnap's notion of quasianalysis (Leitgeb 2007).

As long as the mathematics employed in these applications is used to *model* the entities, or the structure of the entities, or the reasoning we engage in, etc., I have no qualms. But we should not confuse the mathematical entities in a model with the entities being modeled. To give just a simple example: though set-theoretic models of propositions (i.e., ones that treat them as functions from worlds to truth-values or as sets of truth-values) have some interest in so far as they can represent the truth-conditions of, and inferential relations among, the sentences expressing those propositions, we shouldn't identify the propositions expressed with sets or functions. Instead we should try to develop theories of propositions that have these set-theoretic structures as models.

What is the proper role of philosophy of mathematics in relation to logic, foundations of mathematics, the traditional core areas of mathematics, and science?

This question raises a host of further questions, such as: How do we demarcate logic and mathematics and what is the relationship between them? Are there logical objects and how do they differ from mathematical objects? To what extent do logical and mathematical foundations overlap? To begin to answer these questions, let us focus on the questions of how the metaphysical and epistemological foundations of mathematics relate to those of logic and science.

It is important begin by noting that true, ordinary mathematical claims typically occur either (a) in the context of 'natural' or naive mathematics, such as ordinary, naive geometrical claims, ordinary number statements appealing to the natural numbers, and ordinary, naive statements about sets or classes (i.e., extensions of ordinary properties), or (b) explicitly or implicitly in the context of some mathematical theory T. Thus, whenever we attempt to analyze some true mathematical claim, or consider its relationship to logic and science, we must decide whether we have a case of (a) or (b). We shall assume that (a) and (b) are exclusive possibilities, and that any ambiguity must be resolved in one way or the other.[7]

In the theory of abstract objects, we represent true mathematical claims of type (a) as claims about natural mathematical objects definable from the metaphysical and logical primitives of our background theory. In particular:

1. True ordinary and naive geometrical claims are analyzed claims about Platonic Forms, as these are described in Pelletier and Zalta 2000. For example, ordinary claims about the triangle (not made in a context that assumes Euclid's axioms) are analyzed as claims about the Platonic Form of Triangularity (Φ_T), which in object theory is the abstract object that encodes all and only the properties necessarily implied by the property of being a triangle.

2. True ordinary and naive claims about the (natural) numbers are analyzed as claims about the (Fregean) natural numbers developed in Zalta 1999. In that theory, the natural cardinal, the number of Fs ($\#F$), is explicitly defined in terms of metaphysical and logical primitives of object theory. ($\#F$ is the abstract object that encodes all and only the properties G which are in 1–1 correspondence with F on the ordinary objects.) Moreover, this notion can be used to define zero and the *predecessor* relation, and the axioms of Dedekind-Peano number theory can be derived.[8]

3. True ordinary and naive statements about sets, such as ordinary statements about the class of humans not made in the context of set theory, can be analyzed as statements about abstract objects that are *extensions*, as defined in Anderson and Zalta 2004. The extension of F (ϵF) is the abstract object that encodes all and only the properties materially equivalent to F.

By contrast, true mathematical claims of type (b) are represented in object theory in the manner set out in Zalta 2000a. The theorems of each mathematical theory T are imported into the

---

[7] The ideas in this paragraph and the next were first sketched in Zalta 2006.

[8] For the full details, see Zalta 1999. The derivation requires the assumption that Predecessor is a relation, and a modal assumption that guarantees, when n numbers the Gs, that there might have been an concrete object distinct from all the Gs.

theory of abstract objects by (i) prefacing the theory operator "In theory T" to each theorem and (ii) indexing the individual terms and predicates used in T to T. The ordinary claim 'In theory T, ...' is analyzed in object theory as: $T[\lambda y\ \phi^*]$. This latter is an encoding claim for which ϕ is the usual translation of '...' into the encoding-free formulas of classical logic and ϕ^* is just ϕ but with all the terms and predicates of T indexed to T. Thus, mathematical theories are identified as abstract objects that encode propositions by encoding propositional properties of the form $[\lambda y\ \phi^*]$. Now for any primitive or defined individual term κ used in theory T, the object κ_T can be identified as the abstract object that encodes all and only the properties F satisfying the open formula 'In theory T, $F\kappa_T$'. Similarly, in the context of the third-order theory of abstract objects, for any predicate Π appearing in theory T, the relation Π_T can be identified as the abstract relation that encodes all and only the second-order properties $\boldsymbol{F}$ that satisfy the open formula 'In theory T, $\boldsymbol{F}\Pi_T$'.

To make the analysis of the preceding paragraph maximally explicit, here is an example of a natural mathematical object, a theoretical mathematical object, and a theoretical mathematical relation, where the $T \models \psi$ abbreviates $T[\lambda y : \psi]$ in the second and third identities:

$$\Phi_T =_{df} \imath x(A!x\ \&\ \forall F(xF \equiv \Box \forall y(Ty \to Fy))) \qquad (\theta)$$

$$\emptyset_{\text{ZF}} = \imath x(A!x\ \&\ \forall F(xF \equiv \text{ZF} \models F\emptyset_{\text{ZF}})) \qquad (\zeta)$$

$$\in_{\text{ZF}} = \imath \boldsymbol{x}(A!\boldsymbol{x}\ \&\ \forall \boldsymbol{F}(\boldsymbol{x}\boldsymbol{F} \equiv \text{ZF} \models \boldsymbol{F}\in_{\text{ZF}})) \qquad (\eta)$$

(θ) is the explicit definition of the Form of the Triangle (Φ_T) described above. (ζ) is not a definition but a derivable principle that is a consequence of the Reduction Axiom (Zalta 2000a, Section 3).[9] (η) is analogous to (ζ); it is a higher-order object-theoretic principle governing the identity of the membership relation of ZF. (In (η), the variable '$\boldsymbol{x}$' ranges over relations among individuals, and '$\boldsymbol{F}$' ranges over properties of such relations.)

This formal analysis, based on the distinction between (a) natural mathematical objects and (b) theoretical mathematical objects and relations, reveals that logic is more closely related to natural mathematics than it is to theoretical mathematics. The objects of natural mathematics described above look very much like logical objects, for when we compare the definitions of Φ_F, $\#F$,

---

[9] In a separate work, Zalta 2006, (ζ) is described as an instance of a Theoretical Identification Principle (Section 2.4).

and ϵF (introduced in the enumerated paragraphs (1), (2), and (3) above) with the object-theoretic definitions of truth-values, directions, shapes, concepts, possible worlds, impossible worlds, etc., we find that the definitions can all be constructed using logical and metaphysical notions alone.[10] Thus, our metaphysical theory of objects and relations yields both logical objects and natural mathematical objects from its own first principles. By contrast, the analysis of theoretical mathematical objects and relations requires that we import the primitive notions and axioms of (explicit or implicit) mathematical theories into object theory. The synthetic axioms and theorms ϕ of a mathematical theory T are represented in object theory as analytic claims of the form 'In theory T, ϕ^*' (with indexed terms as described above). Each primitive mathematical individual and relation gets associated with a principle that identifies it in terms of an abstract individual or relation that is guaranteed to exist by the principles of the theory of abstract objects. So though the primitive expressions of mathematical theories become imported into object theory, each is accompanied by a principle that offers an analysis of the object or relation it signifies.

This is how I see the philosophical foundations of mathematics as relating to those of logic and the traditional core areas of mathematics. The relationship of the philosophy of mathematics to science is discussed in the answer to the next question.

What do you consider the most neglected topics and/or contributions in late 20th century philosophy of mathematics?

I focus here on two of the most neglected topics from my point of view. The first is the applicability of mathematics to the natural world. There are lots of interesting issues that fall within this topic, such as those that trace back to Wigner's (1960) question about the 'unreasonable effectiveness' of mathematics in science (see Steiner 1998, Colyvan 2001). The basic question I am concerned about is the proper analysis of the language of science. By the 'language of science', I refer not only to the language used in scientific theories (which is often simply mathematical in nature), but also to the language used by scientists themselves as

---

[10] See Anderson and Zalta 2004 for the theory of truth-values, directions, and shapes, Zalta 2000b for the theory of concepts, Zalta 1993 for the theory of possible worlds, and Zalta 1997 for the theory of impossible worlds.

they consider and formulate hypothesis, design experiments, etc. The language of science contains both (1) expressions that refer to concrete, spatiotemporal objects and to the relations among them, as well as (2) expressions for mathematical objects and relations. The philosophers of science who think that science is primarily about building models and representations of objects and processes in the natural world won't face much of a problem when analyzing the language of science, since models for them are essentially set-theoretic structures and they accept set theory (and model theory) as part of the philosophical foundations of science. But I don't accept set theory or model theory as part of the philosophical foundations of science, unless these theories are analyzed in the manner described in the previous sections. The language of science should not be interpreted in terms of (pure or impure) set-theoretic models unless it is explicitly intended to be language about models of the natural world instead of directly about the world itelf. With these distinctions, I can say that my interest lies in the question of *how* to analyze the mathematical expressions in language about the natural world used by scientists in their theories and in their everyday work.

The problem here is this: it may be that a proper understanding of abstract objects and relations (of which mathematical objects and relations form a subdomain) entails that they are entirely defined by our theories of them. We should not conceive of them as non-natural entities made of some Platonic substance, accessible by some special faculty of intuition. We should not use the model of physical objects to understand the mind-independence and objectivity of abstract objects (or mathematical objects), as Linsky and I have argued (1995). But if mathematical objects are defined by our theories of them, what is the relationship between the objects defined by pure mathematical theories and the objects defined by applied versions of those theories (assuming that the applied theories are non-conservative extensions of the pure ones)? To take a simple example, pure ZF is a different theory from ZF + Urelements (say, for example, ZF + {Socrates} or ZF + {Concrete Objects}). So, does the expression '$\{\emptyset\}$' as it appears in pure ZF denote the same object as the expression '$\{\emptyset\}$' in ZF + {Socrates}? If you think that $\{\emptyset\}$ is an object independent of our theories of sets and accessible to some special faculty of intuition, then you will answer this last question with a 'Yes'. But I don't think that a principled metaphysics and systematic epistemology for this conception of abstract and mathematical objects can be sustained. So I see the general question, of how the objects of our pure mathematical theories relate to the objects of those same theories when applied, as defining an issue that needs to be explored. Clearly, it has direct bearing on how we are to understand the applied mathematics appearing in the language of science. And the issue, as we've described it, just touches the tip of the iceberg, since the problems become even harder when we consider the fact that *relations* sometimes used in (the axioms of) scientific theories are essentially just mathematical relations.

A second neglected topic concerns one of the deepest insights that Frege had concerning the way we apprehend mathematical objects, namely, that we apprehend a mathematical object x when we can extract an identity claim about x from general truths about x. Few philosophers have developed an epistemological mechanism for moving from general mathematical truths to identity claims about mathematical objects. But Frege wrote:

> If there are logical objects at all—and the objects of arithmetic are such objects—then there must also be a means of apprehending, or recognizing them. This ... is performed ... by the fundamental law of logic that permits the transformation of an equality holding generally into an equation. (1903, § 147)

The fundamental law Frege was referring to here is Basic Law V, and he assumed that it legitimately turned equalities holding generally (of the form $\forall x[f(x) = g(x)]$) into an equation of the form $\epsilon f = \epsilon g$. (Here I am taking liberties with Frege's notation by assuming that 'ϵ' is an operator on functions f so that ϵf is the course-of-values of the function f.)

Frege was putting his finger on something important here. The question is: how do we transform fundamental logical and mathematical truths into metaphysical (or not purely mathematical) identities in which the expressions denoting mathematical objects and relations constitute one of the terms flanking the identity sign? Such identities are required no matter what our background philosophy of mathematics is, for every philosophy of mathematics needs to precisely state the semantic significance of mathematical terms and predicates. We are obliged to say what the significance is of terms like 'the triangle', 'the number of planets', 'the class of humans', '3', 'π', '$\emptyset$', 'ω', '$\aleph_0$', etc., and predicates and operations such as $<$, $\leq$, $\in$, $\oplus$, etc. And one must give some account of how the significance of these expressions is related to the significance of the sentences in which they appear, no matter whether one is a platonist, structuralist, inferentialist, etc. It is even encumbent on nominalists and fictionalists to give an account of the semantic significance of these expressions in so far as they contribute to the meanings of (false) mathematical sentences.

In the theory of abstract objects, the identities in question result either by explicit definitions or by way of reduction principles, depending on whether the entity in question is a natural mathematical object or a theoretical mathematical object or relation, respectively. Consider our examples (θ), (ζ), and (η) above. These instances of definitions and principles show how general claims and theorems can be transformed into identities of the objects in question. (θ) asserts the identity of the Form of the Triangle, which is defined in terms of the properties F necessarily implied by being a triangle (T). This latter generality ($\Box \forall y (Ty \to Fy)$) is incorporated into the description of the abstract object on the right side of the identity claim. By contrast, (ζ) asserts the identity of the empty set of ZF, which is specified in terms of the properties attributed to the empty set in the theorems of ZF (the theorems of ZF become imported into object theory prefaced by the operator 'In ZF'). Thus, all and only the properties satisfying the formula 'ZF $\models F\emptyset_{ZF}$' are encoded in the empty set of ZF. And similarly for (η), which asserts the identity of the membership relation of ZF: all and only the second-order properties $\boldsymbol{F}$ satisfying the formula 'ZF $\models \boldsymbol{F}\in_{ZF}$' are encoded in the membership relation of ZF.

So this is the means by which the identities of particular mathematical objects and relations can be extracted from the general mathematical claims that govern them. The epistemological significance of this cannot be overstated, for now we simply need the faculties of the understanding and reasoning to become acquainted with mathematical objects. When coupled with the version of neologicism defended in Linsky and Zalta 2006 (on which mathematics becomes reducible to weak third-order logic and analytic truths),[11] it becomes clear that no special faculty (such as intuition) is needed to apprehend mathematical objects and relations or to recognize the truth of mathematical claims. Our work

---

[11] By 'weak' third-order logic, we mean a logic that is no more powerful than first-order logic with separate domains for first-order relations and for second-order relations, and which assumes nothing more than weak comprehension principles for those domains, the smallest models of which require that there are only two first-order relations and four second-order relations.

suggests that theoretical mathematical truths are reducible to (1) the analytic principles of weak third-order logic, (2) an analytic abstraction principle for abstract objects, and (3) analytic truths of the form 'In theory T, ϕ'. As such, only our faculties for understanding language and drawing inferences are required for having knowledge of mathematics. This result, I suggest, forms part of the epistemological foundations of mathematics.

What are the most important open problems in the philosophy of mathematics and what are the prospects for progress?

In the answer to this final question, I describe one important open problem and propose its solution. The open question is: how can we *unify* the apparently divergent views in the philosophy of mathematics such as platonism, nominalism, fictionalism, structuralism, and inferentialism? The point behind this question is to suggest that the philosophy of mathematics would do well to have a theory that unifies these positions to whatever extent possible. Such a unification should explain why there is *disagreement* about the data, for example, over the truth values of mathematical sentences. Some philosophers say that the sentences of mathematics are true, while others say they are false, while still others say they are not truth-apt and so not candidates for truth or falsity, etc. So is it possible to unify the different traditions in the philosophy of mathematics and explain (away) these differences?

I believe it is. If we bring together the results of several pieces of prior research, then platonism, nominalism, fictionalism, and structuralism can be unified. Basically the idea is to interpret the formalism for the theory of abstract objects in different ways. One interpretation yields a form of platonism, another a form of nominalism, yet another a form of fictionalism, and yet another a form of structuralism. The interpretation of the theory of abstract objects as a version of platonism is to be found in Linsky & Zalta 1995 (536–541); as a version of nominalism in Bueno & Zalta 2005 (299–305); as a version of fictionalism in Zalta 1983 (Chapter VI), Colyvan & Zalta 1999 (346–348), and Zalta 2000a (255–256); and as as a version of structuralism in Linsky & Zalta 1995 (545–546). To complete this picture, we sketch how to interpret the formalism as a version of inferentialism.

To see how our analysis of mathematical objects and relations becomes a form of inferentialism, it is important to mention first that inferentialism is the view that the meaning of a mathematical term is to be identified with its inferential role in mathematical discourse. Now reconsider the formal claims (described above) that identify of theoretical mathematical objects in object theory. For example, the empty set of ZF is identified above on line (ζ) as:

$$\imath x(A!x \ \& \ \forall F(xF \equiv \text{ZF} \models F\emptyset_{\text{ZF}}))$$

In other words, the empty set of ZF encodes all and only the properties F that satisfy the condition $\text{ZF} \models F\emptyset_{\text{ZF}}$. As we saw earlier, conditions of this form arise when we import all the theorems of ZF into object theory under the theory operator 'In ZF' and index the terms and predicates to ZF. The particular properties of $\emptyset$ satisfying these conditions are therefore keyed to the theorems of ZF involving the term '$\emptyset$', i.e., for any formula $\phi(\emptyset)$ such that $\vdash_{\text{ZF}} \phi(\emptyset)$, one can use λ-abstraction to produce a formula of the form $\vdash_{\text{ZF}} [\lambda z \ \phi(z)]\emptyset$. This latter then picks out a property that satisfies the condition on properties used in the identification of $\emptyset_{\text{ZF}}$. But the ZF-formulas of the form $\phi(\emptyset)$ such that $\vdash_{\text{ZF}} \phi(\emptyset)$ jointly constitute the inferential role of the term '$\emptyset$' in ZF. Thus, our analysis uses a double abstraction process to abstract out the inferential role of '$\emptyset$' in ZF and objectify it: λ-abstraction on the theorems of ZF involving '$\emptyset$' yield theorems identifying the derivable properties of $\emptyset$ according to the theory, and object abstraction reifies those properties as an object. On this conception, (ζ) identifies the significance of '$\emptyset$' in the theory ZF as its inferential role in that theory.

If we generally apply this conception to all the other theoretical identifications of the objects and relations of mathematical theories, we have an inferentialist interpretation of mathematics. For each term or predicate of mathematical theory T becomes identified with nothing other than the inferential role of that expression in T, something that can be precisely described in the theory of abstract objects. This, then, completes our answer to the open question about the unification of the divergent views in the philosophy of mathematics. The different philosophies of mathematics can now all be seen as different interpretations of the same underlying formalism! Moreover, the disagreement about the data is explained by the fact that ordinary theoretical claims in mathematics become ambiguous. Unadorned claims of mathematics such as '2 is prime', '$\emptyset$ is an element of $\{\emptyset\}$', etc., have exemplification readings (which are false, but which become true when prefaced by the theory operator) as well as encoding readings (which are true).[12] The platonists focus on the true readings to the exclusion of the false, while the nominalists and fictionalists focus on the false readings to the exclusion of the true. The structuralists and inferentialists, meanwhile, focus on the 'incompleteness' of the structures and inferential roles. Heretofore, it was widely thought that no form of classical logic could treat such incomplete objects as the indeterminate elements of mathematical structures (objects defined only by their mathematical properties) and the inferential roles of the terms and predicates of mathematical theories (defined only by the theorems in which they play a part). But it would be an invalid inference to draw such a conclusion in the context of the logic of encoding and the theory of objects.

Bibliography

Anderson, D., and E. Zalta, 2004, 'Frege, Boolos, and Logical Objects', *Journal of Philosophical Logic*, 33/1: 1–26.

Bueno, O., and E. Zalta,, 2005, 'A Nominalist's Dilemma and its Solution', *Philosophia Mathematica*, **13**: 297–307.

Colyvan, M., 2001, 'The Miracle of Applied Mathematics', *Synthese*, 127: 265–277.

Colyvan, M., and E. Zalta, 1999, 'Mathematics: Truth and Fiction?' (Critical Study of Mark Balaguer's *Platonism and Anti-Platonism in Mathematics*), *Philosophia Mathematica*, 7/3: 336-349.

Frege, G., 1891, 'Function and Concept', translated by P. Geach, in P. Geach and M. Black, *Translations from the Philosophical Writings of Gottlob Frege*, Oxford: Blackwell, 1980, pp. 21–41.

Frege, G., 1903, *Grundgesetze der Arithmetik*, Volume II, Jena: Verlag Hermann Pohle.

Leibniz, G., 1690, 'A Study in the Calculus of Real Addition', in *Leibniz: Logical Papers*, edited and translated by G.H.R. Parkinson, Oxford: Clarendon, 1966, pp. 131–144.

---

[12] See Zalta 2000a (Section 6) for a thorough description of how to formulate the exemplification and encoding readings of ordinary mathematical claims. For a more accessible account, see Zalta 2006, 674–678, and 688–691.

Leitgeb, H., 2007, 'A New Analysis of Quasianalysis', *Journal of Philosophical Logic*, 36/2: 181–226.

Linsky, B., and E. Zalta, 1995, 'Naturalized Platonism vs. Platonized Naturalism', *The Journal of Philosophy*, XCII/10: 525–555.

Linsky, B., and E. Zalta, 2006, 'What is Neologicism?', *Bulletin of Symbolic Logic*, 12/1: 60–99.

Pelletier, F.J. and E. Zalta 2000, 'How to Say Goodbye to the Third Man', *Noûs*, 34/2 (June): 165–202.

Shapiro, S., 2004, 'Foundations of Mathematics: Metaphysics, Epistemology, and Structure', *The Philosophical Quarterly*, 54/214: 16–37.

Steiner, M., 1998, *The Applicability of Mathematics as a Philosophical Problem*, Cambridge, MA: Harvard University Press.

Wigner, E., 1960, 'The Unreasonable Effectiveness of Mathematics in the Natural Sciences', *Communications in Pure and Applied Mathematics*, 13/1. New York: Wiley & Sons, Inc.

Zalta, E., 1983, *Abstract Objects: An Introduction to Axiomatic Metaphysics*, Dordrecht: D. Reidel.

———, 1988, *Intensional Logic and the Metaphysics of Intentionality*, Cambridge, MA: MIT Press.

———, 1993, 'Twenty-Five Basic Theorems in Situation and World Theory', *Journal of Philosophical Logic*, 22: 385–428.

———, 'A Classically-Based Theory of Impossible Worlds', *Notre Dame Journal of Formal Logic*, 38/4 (Fall): 640–660 (Special Issue, Graham Priest, Guest Editor).

———, 1999, 'Natural Numbers and Natural Cardinals as Abstract Objects: A Partial Reconstruction of Frege's *Grundgesetze* in Object Theory', *Journal of Philosophical Logic*, 28/6: 619–660.

———, 2000a, 'Neo-Logicism? An Ontological Reduction of Mathematics to Metaphysics', *Erkenntnis*, 53/1-2: 219–265.

———, 2000b, 'A (Leibnizian) Theory of Concepts', *Philosophiegeschichte und logische Analyse/ Logical Analysis and History of Philosophy*, 3: 137–183.

———, 2006, 'Essence and Modality', *Mind*, 115/459 (July 2006): 659–693.

About the Editors

Vincent F. Hendricks is Professor of Formal Philosophy at Roskilde University and member of IIP — Institut Internationale de Philosophie. He is the author of numerous papers and books including *Mainstream and Formal Epistemology*, *Thought$_2$Talk*, *The Convergence of Scientific Knowledge*, *Feisty Fragments*, *Logical Lyrics* and *500 CC: Computer Citations*. Other books include *Self-Reference*, *Proof Theory*, *Probability Theory* and *Knowledge Contributors* and *New Waves in Epistemology*. Editor-in-Chief of *Synthese* and *Synthese Library* he is also the founder of ΦLOG— *The Network for Philosophical Logic and Its Applications* and subject editor for the *Stanford Encyclopedia of Philosophy*.

Hannes Leitgeb holds a joint position as a Professor of Mathematical Logic and Philosophy of Mathematics at the Departments of Philosophy and Mathematics in Bristol. He is a managing editor of *Studia Logica*, an associate editor of *Erkenntnis*, an editor of the *Collected Works of Rudolf Carnap*, and a subject editor in Philosophy of Mathematics for the *Stanford Encyclopedia of Philosophy*. He has also (co-)edited special issues of *Synthese* and *Studia Logica*. He is the author of numerous papers and books on logic, philosophy of mathematics, and epistemology.

About Philosophy of Mathematics

Philosophy of Mathematics is a collection of short interviews based on 5 questions presented to some of the most influential and prominent scholars in this field. We hear their views aim, scope, use, the future direction and how their work fits in these respects.

Philosophy of Mathematics collects together answers on 5 provocative questions by many of the leading contemporary figures in Philosophy and Mathematics, two of the most fundamental and widely applicable intellectual skills. The collection contains ample amount of interesting considerations, far beyond what one finds reflected in standard texts and together they show that one can have surprising, sometimes tortured, but often highly productive relationships between Philosophy and Mathematics. In my opinion, this book affords pleasure to the reader.

— Leo Esakia

WWW.PHIL-MATH.ORG
© 2008   Vincent F. Hendricks & Hannes Leitgeb

Index

Abramsky, S., 40
abstractionism, 223
academic freedom, 155
Ackermann, W., 175
Aczel, P., 242
adjudication, 103
AI, 35
algebra, 7, 46, 75, 89, 106, 175
    Boolean, 20
    computational, 7
    Heyting, 20
algorithm, 296
analysis, 6, 29, 83, 89, 148, 158, 211, 235, 278
    constructive, 46, 156, 195
    predicative, 156
Anderson, A.R., 211, 251
anti-realism, 302
Appel, K., 295
arealism, 177
Aristotle, 19, 261
arithmetic, 63, 122, 176, 223, 235, 236, 278, 290, 292, 315
    Heyting, 126, 185
    intuitionistic, 223
    NFA, 125
    Peano, 2, 39, 91, 108, 120, 159, 187
    primitive recursive, 149, 180, 257
    Robinson, 270
    second-order, 6, 253
Asimov, I., 1

atomism, 19
Australian National University, 75
automated theorem prover, 244
autonomy condition, 123
Avigad, J., 1, 202, 296
Awodey, S., 11
axiom, 70, 177, 178
    abstraction, 70
    extensionality, 70
    of choice, 29, 75, 151, 152, 167, 225
    of infinity, 158, 175, 178, 213, 294
    of separation, 38
    reduction, 319
    replacement, 146, 151
axiomatic field theory, 87
axiomatics, 6, 8, 167, 236, 244
axiomatization, 167, 191
Ayer, A.J., 87

Bacon, F., 102
Baire, R.-L., 152
Barendregt, H., 266
Barwise, J., 194, 267
Basic Law V, 63, 223, 322
Bateman, H., 116
Beals, R., 211
Beck, L.W., 249
Becker, O., 196
belief revision theory, 36
Bell, E.T., 116
Bell, J., 15, 149
Bell, J.S., 154, 155
Belnap, N., 292

Benacerraf, P., 30, 140, 200, 213, 305
Benthem, J.v., 29
Berkeley, G., 16, 70, 106, 301
Bernays, P., 186, 206, 236, 240, 254
Beth, E.W., 32
biology, 77, 178
Bishop, E., 45, 127, 157, 186, 195, 254, 258
bivalence, 21, 22, 280
Bliedtner, J., 184
Bolzano, B., 19, 35, 36, 153, 260
Borel, E., 152
Bostock, D., 277
Bourbaki, 16, 33, 233
brain, 25, 42, 180
Brandenburger, A., 177
Bridges, D., 45, 157
Bronowski, J., 87
Brouwer, L.E.J., 3, 19, 27, 32, 46, 59, 127, 153, 156, 205, 207, 216, 241, 250, 258
Buchholz, W., 2
Burgess, J.P., 55, 220, 270

calculemus, 8
calculus, 19, 80, 176
CalTech, 116, 175
Cantor, G., 1, 23, 122, 127, 153, 177, 183, 235, 256, 260
cardinals, 175, 191, 222, 273
 inaccessible, 163
 large, 125, 127, 173, 197
 axioms, 294
 Woodin, 176
Carnap, R., 11, 36, 115, 158, 170, 207, 216, 317
Carnegie Mellon University, 7

Case Western Reserve University, 220
Case, J., 220
categoricity, 222, 290, 297
category theory, 11–13, 18–20, 23, 153, 158, 178, 202, 233, 272
Cauchy, A.L., 93
Chang, C.C., 175
Charles University, 173
Chihara, C., 51, 213
Church's thesis, 214, 221, 238, 303
Church, A., 39, 118, 187, 238
Church-Kleene recursive ordinal notation, 121
Churchland, P., 93
Clark, P., 277
Clay Mathematics Institute, 130
coalition, 179
cognitive science, 42
Cohen, P., 20, 153, 169, 173, 188, 278, 297
cohomology, 202
Colyvan, M., 75
compactness, 4, 187, 222
completeness, 12, 187, 297
comprehension, 38, 185, 186, 253, 292
computability, 20, 27, 89, 148, 221, 235, 239
computational complexity, 36, 178
computational ergodic theory, 7
computational number theory, 7
computational reduction, 237
computer science, 2, 31, 188, 202, 220, 240, 271, 272

concept addition, 316
concept inclusion, 316
Connes, A., 91
consistency, 285
constructivism, 46, 47, 54, 127, 148, 220
contextualism, 228
continuity, 20, 24, 26, 56, 149, 150, 166
continuum hypothesis, 129, 140, 173, 177, 187, 191, 258, 284, 290
Coquand, T., 6
Corcoran, J., 220
Cornell University, 251
Courant, R., 249
creative subject, 188
Curry, H., 266
Curry, P., 30
cut-elimination, 253

Dalen, D.v., 266
Dartmouth College, 277
Davies, E.B., 87
De Groot, J., 29
decision theory, 37, 76
Dedekind, R., 153, 183, 235, 256
deduction, 32, 33, 37, 41, 213
definability, 8, 33, 34, 120, 128, 187, 221, 238, 292
deflationism, 281
demonstration, 103
derivability condition, 243
Descartes, R., 3, 16, 19, 198, 301
design argument, 77
Detlefsen, M., 101, 202
Dewey, J., 115
diagonalization, 289
Dialectica interpretation, 184, 252
dialetheism, 281, 286

differentiability, 20
Dijkgraaf, R., 42
Dirichlet, J.P.G., 235
discovery, 103, 201
Dreben, B., 145, 212
Dretske, F., 36
Dummett, M., 59, 137, 141, 148, 161, 259, 301, 302, 307, 310
Durant, W., 249

economics, 179, 227
  mathematical, 179
Eddington, A.S., 115
Einstein, A., 115, 157, 214
eliminative materialism, 93
epistemology, 10, 36, 41, 76, 137, 144, 146, 192, 200, 214, 216, 222, 228, 279, 285
equicardinality, 166
equilibrium, 179
Erlanger Programm, 34
Etchemendy, J., 194, 267
Euclid, 3, 104
Euler, L., 3
exceptionalism, 98
existence, 37, 178, 303, 313
experimentation, 178
externalism, 228

Fajardo, S., 176
Feferman, A.B., 117, 175
Feferman, S., 115, 151, 156, 159, 175, 194, 234, 252, 258, 290
Feynman, R., 233
fiction, 27
fictionalism, 127, 282, 303, 324
Field, C., 244
Field, H., 55, 75, 152, 160, 213, 216, 279, 282, 305

Fine, K., 270, 272
fine-tuning argument, 77
finitism, 123, 126, 127, 241, 256–258, 271, 282
Fitch, F., 211, 250
Fleischhacker, L., 266
Føllesdal, D., 212
forcing, 5, 20, 173, 188, 294, 297
formal learning theory, 36
formalism, 11, 17, 106, 127
formality, 235
formalization, 7, 191, 193
foundational reduction, 237
foundationalism, 41, 228
Four colour problem, 295
Fourier transform, 94
Frege, G., 11, 19, 22, 38, 40, 60, 127, 153, 158, 165, 167, 207, 212, 213, 216, 223, 234, 236, 256, 257, 270, 281, 289, 301, 306, 322
Friedman, H., 8, 147, 151, 188, 220, 253
Friedman, M., 257
functors, 19, 178

Galileo, G., 24
game theory, 37
Gandy, R., 235
Ganea, M., 270
Garfield, J., 76
Gauss, C.F., 3, 92, 112, 165
Geach, P., 270
Gentzen, G., 122, 237
geometry, 19, 199, 236, 278
    differential, 96, 148
    Euclidean, 229, 269, 307
Gerhardy, P., 7
Giaquinto, M., 202
Gillies, D., 88
Giorello, G., 194

Glashow, S., 155
Gödel, K., 15, 40, 91, 121, 153, 177, 207, 214, 238, 253, 256, 278
Goldberg, S., 250
Gonthier, G., 7, 295, 296
Goodman, N., 147, 214, 220
Groenendijk, J., 268
Grotemeyer, P., 233
Grünbaum, A., 250
Gupta, A., 292

Hailperin, T., 250
Hájek, A., 77
Hale, B., 137, 216, 306
Hales, T., 7, 95, 296
Halmos, P., 252
halting problem, 215
Hameroff, S.R., 91
Harrington, L., 220, 278
Harris, M., 150
Harvard University, 145, 212
Hawking, S., 157
Hegel, G.W.F., 16
Heijenoort, J.v., 30
Hellman, G., 55, 145, 159, 213, 216
Henkin, L., 118, 251
Herbrand, J., 237
Hesse, H., 47
Heyting, A., 30, 153, 216
hierarchialism, 286
Hilbert's program, 2, 34, 38, 108, 205, 234, 237, 254, 257
Hilbert, D., 13, 105, 107, 127, 130, 155, 165, 167, 175, 183, 186, 206, 207, 213, 235, 236, 240, 254, 256
Hilbert-Ackermann lemma, 193
Hilbert-Bernays derivability, 120

Hintikka, J., 30, 165
Hofstader, D., 155
Hollak, J., 266
Hoyle, F., 15
Hume's principle, 60, 159, 223, 270, 297, 304, 310
Hume, D., 16
Husserl, E., 165, 194, 262
Huxley, J., 87
hyperfinite set, 176, 178

impredicativity, 123
incompleteness, 13, 119, 120, 326
  arithmetical, 121
indispensability thesis, 76, 126, 128, 199
induction, 4, 170, 290
inferentialism, 324
infinitesimals, 176
infinity, 127, 146, 159, 166, 196, 214, 260, 289
information theory, 36, 37
Institute for Advanced Study, Princeton, 116, 253
instrumentalism, 55, 127, 152
integral, 176
integration challenge, 305
intensionality, 221
intuitionism, 19, 59, 153, 156, 159, 161, 220, 241, 250, 257, 259
invariant, 296
invention, 103
Islamic fundamentalism, 40

James, W., 16
Jeans, J., 115
Jech, T., 173, 220
Jensen's covering lemma, 294
jurisprudence, 103
justification, 8, 201, 228

Kakutani, S., 211
Kant, I., 16, 19, 32, 98, 206, 217, 230, 234, 254, 261, 278, 301
Kaulbach, F., 234
Kearns, J., 220
Keisler, H.J., 175
Kelley, J., 15
Kelly, K., 36
Kepler conjecture, 7, 95
Ketonen, J., 2
King's College, London, 88
Kino, A., 220
Kleene, S.C., 118, 206, 251
Klein, F., 34
Knorr, W., 195
knot theory, 202
Kochen, S., 154
Kohlenbach, U., 6, 183, 254
König's lemma, 166, 185
Körner, S., 137
Korzybski, A., 115
Kracht, M., 272
Krantz, D., 211
Kreisel, G., 118, 119, 206, 230, 234, 252
Kripke, S., 187, 266, 291
Kronecker, L., 46, 235, 254, 258

Lakatos, I., 35, 88
Lakoff, G., 25
Lane, M., 11, 13
language, 27, 106, 222, 260, 286, 291
  arithmetic, 4
  first-order, 4
  formal, 176
    semantics, 176
    syntax, 176
  higher-order, 4
  of science, 320
  set theory, 4

Lawvere, W., 18
Lebesgue, H., 152
Lehigh University, 249
Leibniz, G.W.F., 3, 16, 19, 198, 261, 301, 316
Leitgeb, H., 41
Lewy, C., 301
limit, 176
Linsky, L., 11, 321
Löb, M., 30
Locke, J., 16
logic, 2, 29, 52, 129, 150, 166, 175, 184, 197, 205, 216, 238, 240, 307
   basic, 251
   bivalent, 21
   classical, 80, 185, 286, 307
   combinatorial, 251
   constructive, 11
   dynamic, 36
   dynamic predicate, 268
   epistemic, 36
   extensional, 31
   first-order, 37, 166, 176, 222, 234, 268
   higher-order, 12, 37, 167, 222
   IF, 166, 169
   intensional, 31
   intuitionistic, 18, 33, 39, 222, 307
   mathematical, 2, 5, 8, 12, 19, 30, 107, 175, 183, 194, 208, 235, 290
   mathematics, 187
   modal, 30, 31, 176, 178
   non-classical, 170
   ordinal, 122
   paraconsistent, 80, 81
   philosophical, 76
   propositional, 30
   provability, 278
   quantum, 20
   second-order, 169, 297
   temporal, 36
   trivalent, 116
logical atomism, 306
logicism, 11, 60, 127, 158, 257
Luckhardt, H., 184
Łukasiewicz, J., 116
Lusin, S., 294

Mackey, G., 89
Maddy, P., 75, 177, 191, 201, 214
Malament, D., 191
Malcolm, N., 17
Mancosu, P., 193
Manders, K., 202
Maoist mathematics, 152
Markov principle, 185
Martin-Löf, P., 127, 141, 254, 296
Masterman, M., 205
mathematical activity, 293
mathematical concept, 25, 192, 244, 293
mathematical existence, 177
mathematical explanation, 202
mathematical intuition, 176, 178, 179
mathematical knowledge, 8, 24, 33, 138, 207, 209, 216, 314
mathematical object, 33, 48, 75, 177, 180, 230, 242, 292
mathematical practice, 129
mathematical theory, 314, 325
mathematics, 1, 10, 13, 15, 49, 129, 177, 271
   applicability, 24, 25, 75, 141
   beauty of, 29
   classical, 254, 257

338 Index

computational, 23
constructive, 7, 23, 39, 178, 184, 186, 194, 195, 253, 257
  epistemology of, 306, 315
  for philosophy, 303
  foundations of, 30, 38, 52, 127, 150, 175, 197, 205, 208, 240, 255, 314
  history of, 109, 227
  pure, 80, 301
  social activity, 41
  sociology of, 90
mathematizing, 256
McCarthy, J., 35
McLarty, C., 202
meaning, 176
membership relation, 168
metamathematics, 36, 120, 279
metaphysics, 137, 146, 192, 200, 222, 314
  experimental, 155
Meyer, B., 81
Mill, J.S., 217, 305
Millennium Prize Problems, 130
model theory, 8, 30, 38, 63, 169, 175, 176, 194, 220, 222, 279, 321
  finite, 176
monadism, 19
Moore, G.E., 16, 304
Mortensen, C., 81
Mostowski, A., 117
Myhill, J., 220, 250
mystery, 175

natural science, 314
naturalism, 200, 279, 293
  mathematical, 127
  scientific, 127
neo-logicism, 216

Nerode, A., 252
Neumann, J.v., 256
New V, 224
Newton, I., 3, 198
nominalism, 55, 127, 303, 324
non-standard analysis, 176
non-standard universes, 178
number, 145, 269, 270, 273
Nuñez, R., 25
Nutter, J.T., 222

objectivity, 110
Ohio State University, 219
Oppenheimer, R., 116
ordinals, 273
Orevkov, V., 188

paradox, 65, 259, 283
  Burali-Forti, 284
  iterated belief, 177
  liar, 65, 69, 267
  Russell, 3, 71, 158, 191, 214, 279, 289
  Russell, 176
  Skolem, 214
  Zeno, 18
Paris, J.P., 278
Parsons, C., 205, 212
particle physics, 79
Pascal's Wager, 77
Peacocke, C., 305
Pears, D., 137
Peirce, C.S., 36
Penrose, R., 91, 95, 157
Perelman, C., 38
phenomenology, 196, 262
Philip Marlowe, 28
philosophy, 2, 8, 17, 29, 31, 207, 214, 224
  of language, 137, 228
  of logic, 197
  of mathematics, 37, 76, 80, 176, 200, 262

mainstream, 23
of science, 208
physics, 179, 191
planetary system, 38
Plato, 16, 32, 91, 98, 102, 209, 249
Platonism, 17, 91, 96, 212, 255, 262, 303, 321, 324
pluralism, 146, 147
Poincaré conjecture, 27
Poincaré, H., 123, 198, 240
Popper, K., 19, 88, 90, 93, 193, 266
population ecology, 80
Post's problem, 223
Prawitz, D., 141
predicativity, 124, 127, 156, 186
Priest, G., 76, 81, 281, 286
probability theory, 19, 37, 178
Proclus, 104
proof, 108, 177, 228, 295
    consistency, 237, 253, 294
    constructive, 82, 184
    mathematical, 9, 48
    non-constructive, 82
    Zorn-free, 267
proof assistant, 7
proof mining, 6
proof theory, 5–8, 30, 38, 107, 184, 194, 212, 220, 222, 254, 279, 292
proof, unwinding, 6, 184
provability, 2, 120, 124, 128, 198, 221, 297
Pryor, J., 304
pseudospectral theory, 89
Putnam, H., 30, 76, 145, 301

quantification, 37, 54, 161, 166, 222, 308
quantum field theory, 87, 227
quantum mechanics, 154, 155
quantum theory, 19, 20, 96, 154
quasi-empiricism, 127
quasianalysis, 317
Quine, W.V., 54, 76, 79, 145, 167, 205, 207, 211, 215, 216, 230, 261, 266, 277, 301

Ramsey, F., 207
randomness, 296
Rathjen, M., 254
realism, 17, 127, 177, 302
realizability interpretation, 185
reasoning, 4, 42, 103, 109, 168, 173
    diagrammatic, 202
recursion theory, 19, 38, 117, 194, 209, 220, 234
reducibility, 314
reductionism, 98, 303
reflection schema, 121, 122
Reichenbach, H., 16
relativism, 92
relativity theory, 19
remediation problem, 71
Resnik, M., 76, 211, 222
revisionism, 225, 226
Riemann zeta function, 296
Riemann, G., 165, 256
Robbins, H., 249
Robinson, A., 176, 178
Rosen, G., 55
Ross, A., 219
Russell, B., 9, 11, 16, 19, 64, 79, 87, 115, 145, 165, 167, 207, 216, 236, 266, 281, 285, 289, 301, 306, 310
Ryle, G., 137

Saunders, S., 13
Sawyer, W.W., 45

Schiffer, S., 304
Schirn, M., 160
Scholz, H., 234
Schopenhauer, A., 16
Schütte, K., 123, 206
Schwarz, W., 184
science, 80, 257, 293, 301
scientific knowledge, 307
Scott, D., 118, 285
self-reference, 292
set theory, 13, 19, 23, 80, 151, 159, 168, 173, 175, 176, 178, 191, 194, 209, 220, 234, 256, 279, 285, 290, 297, 302, 315
    constructive, 46
    cumulative hierarchy, 178
    GCH, 280
    NBG, 38
    ZF, 38, 160, 188, 217, 222, 321
    ZFC, 79, 92, 108, 160, 224, 258, 280, 292, 294
Shankar, N., 6
Shapiro, S., 219, 306
Sherlock Holmes, 28
Shimony, A., 155
Sieg, W., 233
Silver, J., 2
Simpson, S.G., 53
Skolem, T., 56, 309
Smart, J., 76
smoothness, 149
Smoryński, C., 266
Sober, E., 77
social science, 178
Socrates, 102
Solovay, R., 2, 173
soundness, 237, 285
space, 32
    Euclidean, 34
    Hilbert, 20, 24, 157, 293
    proximity, 20, 26
    Sobolov, 87
Spinoza, B.d., 16, 158
Stanford University, 119, 194, 234, 252
State University of Milan, 194
Statman, R., 188
Steen, S.W.P., 205
Stein, H., 11
Steiner, M., 147
Stokhof, M., 268
string theory, 178
structuralism, 11, 127, 222, 324
    modal, 127
structure, 3, 23, 34, 168, 169, 290
    tame, 176
Sullivan, K., 56
synechism, 19
Syntax, 271

Taine, J., 116
Tait, W., 249
Takeuti, G., 253
Tappenden, J., 202
Tarski curse, 166
Tarski, A., 19, 36, 117, 119, 137, 153, 166, 175, 187, 252, 267, 284
Technical University, Berlin, 233
Tennant, N., 277, 283
The Journal of Symbolic Logic, 118
theology, 31
theorem, 108
    Bolzano-Weierstrass, 149
    Cauchy's integral, 83
    classification, 95
    deduction, 30

definability, 32
Feit-Thompson, 7
Fermat, 27
Frege, 22, 23, 61, 62, 223, 304
Gleason, 157
Gödel's incompleteness, 3, 13, 18, 38, 95, 109, 117, 145, 167, 214, 237, 258, 293, 303
Green, 82
Jordan curve, 7
Lindström, 222
Löwenheim-Skolem, 212, 222
Paris-Harrington, 2
residue, 83
Skolem, 309
Tarski's, 291
Tychonoff, 29
uniqueness, 185
Thomas, W., 212
Thompson, F., 117
Thomson, J., 205
topology, 15, 29, 34, 46, 75, 199
  algebraic, 11, 227
topos theory, 18, 20
Towsner, H., 7
transfinite induction, 122
translatability, 314
trigonometry, 83
Troelstra, A., 30, 184
truth, 19, 32, 66, 151, 166, 279, 282, 290, 291
Turing, A., 39, 40, 121, 187, 235, 239
type theory, 4, 23, 39, 64, 127, 158, 254, 285
typical ambiguity, 162

UCLA, 116, 175
uncertainty, 179
undecidability, 278
unification, 324
University of Bristol, 137
University of California, Berkeley, 2, 117
University of Cambridge, 15, 205, 301
University of Chicago, 11, 117
University of Delaware, 249
University of Marburg, 11
University of Oxford, 137, 146
University of Pittsburgh, 250
University of St. Andrews, 277
University of Twente, 265
Urquhart, A., 203
USC, 175

variable, 271
Velde, E.v.d., 266
verification, formal, 7
Vesley, R., 220
Visser, A., 265
Vopenka, P., 173

Wainer, S., 2
Wallis, J., 110
Wang, H., 54, 118, 206, 212
Weierstrass, K., 153, 235
Weinberger, O., 38
Weir, A., 224
Welch, P., 289
Wetterling, W.W.E., 265
Weyl, H., 126, 156, 196, 240, 258
Wheeler, J.A., 154
Whitehead, A.N., 145, 167, 207, 236
Wigner, E., 24
Wilder, R., 53
Wiles, J., 304
Wilhelms-University, Münster, 234
Williamson, T., 303

Wittgenstein, L., 11, 17, 22,
    194, 216, 243, 259,
    293, 301, 302, 306
Woodin, H., 294, 297, 298
World 2, 93
World 3, 93
Wright, C., 61, 138, 143, 216,
    277, 301

Yale University, 211, 250

Zalta, E.N., 313
Zermelo, E., 153, 236, 256,
    281, 295
Zorn's lemma, 45
Zucker, J., 266

www.ingramcontent.com/pod-product-compliance
Lightning Source LLC
Chambersburg PA
CBHW021830220426
43663CB00005B/195